Biology: A Critical-Thinking Approach

Anton E. Lawson

Laboratory Manual

Addison-Wesley Publishing Company
Menlo Park, California • Reading, Massachusetts • New York
Don Mills, Ontario • Wokingham, England • Amsterdam • Bonn
Paris • Milan • Madrid • Sydney • Singapore • Tokyo
Seoul • Taipei • Mexico City • San Juan

Editorial and Production: Navta Associates, Inc.
This book is published by Innovative Learning™, an imprint of Addison-Wesley's
Alternative Publishing Group.

ISBN 0-201-46730-5
1 2 3 4 5 6 7 8 9 - ML - 97 96 95 94 93

This book is printed
on recycled paper.

Contents

Preface to the Student iv
Lab Report Guidelines v
Laboratory Safety vii
Investigation 1: What Caused the Water to Rise? 1
Investigation 2: How Are Living Things Classified? 3
Investigation 3: How Do Characteristics Vary? 6
   Reasoning Module 1: Probability 11
Investigation 4: Which Characteristics Are Linked? 15
   Reasoning Module 2: Correlation 21
Investigation 5: What Can Be Learned From Skulls? 31
Investigation 6: What Causes Changes in Heart Rate? 34
Investigation 7: In What Ways Can Isopods Sense Their Environment? 36
   Reasoning Module 3: Causal Relationships 40
Investigation 8: What Is Inside a Frog? 47
Investigation 9: What Do Living and Nonliving Things Look Like Under the Microscope? 49
Investigation 10: How Does Cell Structure Relate to Function? 51
Investigation 11: What Is Inside Cells? 53
Investigation 12: What Gas Do Growing Yeast Cells Produce? 56
   Reasoning Module 4: Controlling Variables 59
Investigation 13: How Do Multicellular Organisms Grow? 65
   Reasoning Module 5: Doing Science 69
Investigation 14: What Happens During the Development of Chicken Eggs? 77
Investigation 15: What Are Foods and Beverages Made Of? 84
Investigation 16: What Is the Function of Saliva? 88
Investigation 17: What Happens to Molecules During Chemical Breakdown? 91
Investigation 18: How Do Molecules Pass Into and Out of Cells? 94
Investigation 19: What Is the Molecular Structure of the Genetic Material? 99
Investigation 20: Where Will Brine Shrimp Eggs Hatch? 101
   Reasoning Module 6: Proportional Relationships 104
Investigation 21: What Is the Structure and Function of Flowers? 111
Investigation 22: What Is the Function of Fruits? 114
Investigation 23: Where Do Seeds Get Energy? 117
Investigation 24: What Variables Affect Plant Growth? 119
Investigation 25: What in the Air Do Plants Need to Grow? 121
Investigation 26: How Do Different Parts of Leaves Function? 124
Investigation 27: What Colors of Light Are Used During Photosynthesis? 129
Investigation 28: Is Chlorophyll Necessary for Photosynthesis? 133
Investigation 29: What Causes Water to Rise in Plants? 137
Investigation 30: What Could It Be? 140
Investigation 31: What Happens to Dead Organisms? 143
Investigation 32: What Is the Pattern of Population Growth and Decline? 146
Investigation 33: What Causes Population Size to Fluctuate? 150
Investigation 34: How Do Organisms Interact With Their Environment? 155
Investigation 35: What Changes Occur in a Temporary Pond? 160
Investigation 36: What Changes Have Occurred in Organisms Through Time? . . . 162
Investigation 37: Have Humans Been on Earth a Long Time? 166
Investigation 38: How Do Species Adapt to Environments? 169

Preface to the Student

This course has been designed to improve your ability to think scientifically by investigating living things and their surroundings. Biological investigations have two key questions. The first is, *What is happening?* The second is, *What caused it to happen?* The first question calls for an adequate description of objects, events, and situations, while the second, more difficult question calls for an explanation.

Unfortunately, many high-school students reach conclusions about explanations in the way diagrammed below:

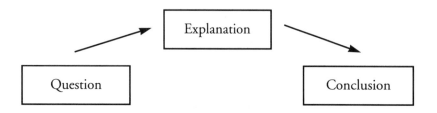

A question is raised, a single explanation (answer) pops into their heads, and they immediately conclude that it is correct without generating and attempting to test any alternatives. This course will help you think more scientifically, like this:

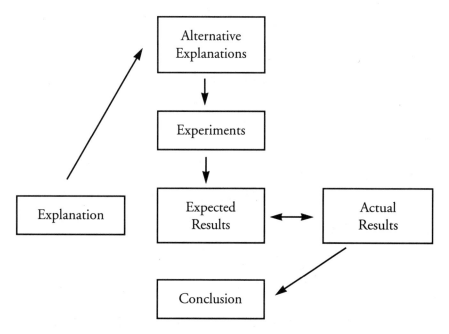

In other words, many possible explanations are proposed and conclusions are not drawn until all the possibilities have been tested thoroughly by comparing expected results of experiments with actual results of those experiments.

Keep in mind that doing science means you must do two things. You must keep alert to discover what is happening, and you must always ask why it is happening. Not *why—What is the purpose?*—but *why—What is the cause?* In brief, your job is to do the following:

- make observations and look for patterns
- raise questions about causes
- create a number of possible explanations (referred to as alternative hypotheses)
- design and conduct experiments to test possible explanations
- collect, organize, and analyze data
- draw reasonable conclusions from the evidence
- reflect back on what you have done and heard and what it means

By the end of the year, you should be quite good at doing these things. You should know quite a lot more about the nature of life on Earth, and you should also be a lot better at thinking like a scientist. The scientific thinking patterns you learn here will serve you well in any future occupation you choose.

Lab Report Guidelines

During the year, you will be doing a number of investigations. For some of these, you may be required to submit a detailed lab report based on your investigation. Following is information to help you with the format of the lab report.

Causal Question

In this section, you should state the causal question you are addressing. You should include an introduction consisting of any background information your reader might need and a discussion of why the question is important or interesting biologically. Essentially, the question you address has been supplied to you in the titles of the investigations, but you probably will find it necessary to include additional questions. For example, in addition to asking the causal question, *What causes molecules to pass into and out of cells?*, you might also want to ask a descriptive question such as, *How does the concentration of molecules affect their movement into and out of cells?*

Alternative Hypotheses (Explanations)

In this section, you will present the alternative hypotheses (at least two) that you will be testing. Be sure your proposed explanations can be tested with the facilities at your disposal. For example, suppose you hypothesize that water rises in plants because little pumps are in the roots, and/or because one-way valves are in the stems. These are reasonable hypotheses. To test them, you must be able to imagine and conduct an experiment that yields expected results (predictions) and actual results (your data). Your hypotheses must be ones that are testable.

Experimental Procedure

This section will describe what you did to test your proposed explanations. You should include a diagram of your setup and enough verbal description so that someone unfamiliar with your experiment could repeat it. State your independent variable (the factor that you manipulate) and your dependent variable (what you are measuring) for each experimental design. Be sure that you have designed controlled experiments and that you have enough data to differentiate random variations from "real" variations. You may need to repeat your experiment several times.

Expected Results (Predictions)

Your expected results (predictions) are derived from your proposed explanations and your experimental design. To generate a prediction, one assumes for the purpose of investigation that the explanation is correct. The prediction may be stated as part of an *If-then* statement. The *if* portion is essentially a restatement of your proposed explanation. An explicit *and* portion represents your experiment, and a *then* portion states the results you expect to find if your proposed explanation is correct.

Example: *If* pumps that allow water to rise are in the roots of plants (proposed explanation) and you cut off the roots of a group of experimental plants while not cutting off the roots of a second group of plants (experiment), *then* you would expect that those plants with intact roots would show a greater water rise than those plants without roots (prediction).

Important Note: The *If-then* statement is simply a convenient way to illustrate the relationship among a proposed explanation, an experiment, and the expected results. A hypothesis does not have to include the word *if,* and a prediction does not have to start with *then.* You cannot differentiate hypotheses from predictions just by looking for these cue words because they are frequently omitted. You must understand the difference between them.

Actual Results

In this section, you will present the actual results of your experiment. Your data should be quantitative and should be presented in tables and/or graphs. Be very careful to label clearly the axes on graphs and the columns and rows on tables.

Discussion and Conclusion

In this section, you will identify trends in your data and discuss whether these trends agree or disagree with predictions derived from your proposed explanations. You may discuss any qualitative observations, explain any irregular or abnormal results, and suggest possible improvements in your experimental design. In your conclusion, you should decide whether to accept or reject your hypotheses based on the results of your experiments. Do your actual results agree with your expected results? If so, then the hypotheses have been supported. If you decide to reject proposed explanations, you may be able to suggest additional explanations at this time. Do your results suggest any further investigations of interest? If so, briefly discuss them.

Laboratory Safety

At the start of the year your teacher will inform you of general safety rules and the location of safety equipment such as fire extinguishers, first-aid kits, and eyewash fountains, if available. Review the Safety Guidelines below. Be aware of local laws and regulations regarding safety and the use and disposal of hazardous chemicals. Chemicals should be clearly labeled and kept in a locked storage area accessible only under the teacher's direction. Be aware of your school's policy for treating and reporting injuries.

Safety Guidelines

 Heated Test Tubes — Do not point the open end of a heated test tube toward anyone. Hot materials may shoot out.

Glassware — Do not handle broken glass. Do not force glass tubing into a rubber stopper.

Mixing Acid in Water — To to avoid a possible explosion, do not pour water into acid. Instead slowly pour acid into water while mixing thoroughly.

Sharp Edges — Be careful when using sharp dissecting tools. Make certain that the specimen is securely pinned to a flat surface. Handle coverslips with care.

Burner and Flames — Do not heat alcohol or other flammable materials over open flames. Keep long hair tied back when using burners or flames.

Fumes — Use your hand to waft fumes toward your nose rather than sniffing directly.

Flammable Materials — Handle with care. Follow specific instructions as directed by your teacher.

Poisons and Toxic Materials — Do not taste any materials used in the laboratory unless specifically directed to do so by your teacher.

Turn Off and Unplug — Turn off water faucets, hot plates, and burners and unplug electrical cords after use.

Dangling Cords — Wrap microscope cord so that it will not be tripped over.

 Eye Safety — Wear safety goggles when working with chemicals, open flame, or toxic materials.

 Clothing Protection — Wear your laboratory apron to protect clothing from stains or damage.

 Plants — Do not taste any plant or plant parts. Wash hands with soap and water after handling plants.

 Animals — Handle live animals with care. Be sure animals are given proper food, water, and living space. Never cause pain, discomfort, or injury to an animal.

First Aid

Injuries should be immediately reported to your teacher. Know where the first-aid kit and supplies such as baking soda and boric acid are kept. Consult the following table for responses to specific injuries.

Type of Injury	First Aid
Burns	Flush immediately with cold water.
Cuts and Bruises	Follow instructions in kit. Squeeze puncture wounds to cause some bleeding. Press gauze on minor cuts to stop bleeding. Apply cold compresses to bruises to reduce swelling.
Eye(s)	Flush eye(s) immediately with water for several minutes. Do not rub eye if it contains a foreign object.
Poisoning	Note the specific substance and immediately inform your teacher.
Spills on Skin	Flush with water if substance is not a strong acid or base. Strong acid – Apply baking soda. Strong base – Apply boric acid.

Investigation 1 *What Caused the Water to Rise?*

Introduction

Often things seem simpler at first glance than they really are. On closer examination, the complexity and mystery become more apparent. Discovering and solving these mysteries can be enjoyable and more satisfying than looking for answers in books or asking people who claim to know better than you. One way to search for your own answers is called science, and it can be fun. You are going to do some now.

Objectives

1. To become curious about natural phenomena.

2. To become aware that science is activity that involves generating and testing possible explanations (alternative hypotheses) to arrive at conclusions.

Materials

water   rubber tubing
aluminum pan   cylinders (open at one end)
birthday candles   jars (of various shapes and sizes)
modeling clay   beakers and/or test tubes
matches   syringes

Procedure

1. Select a partner and obtain the materials.

2. Pour some water into the pan. Stand a candle in the pan; use clay for support.

3. Light the candle. Put a cylinder, jar, or beaker over the candle so that it covers the candle and sits in the water. **Caution: Be careful when using an open flame. Tie back loose hair and clothing.**

4. What happened?

   _____

   _____

5. What questions are raised?

   _____

   _____

   _____

   _____

6. What possible explanations (alternative hypotheses) can you suggest for what happened?

_____

_____

_____

_____

_____

7. Repeat your experiment in a variety of ways (e.g., change the number of candles, amount of water, type of cylinder) to see whether you obtain similar or different results. Do your results support or contradict your ideas in step 6? Explain.

_____

_____

_____

_____

_____

_____

_____

_____

_____

_____

Investigation 2 *How Are Living Things Classified?*

Introduction

People have long been impressed with the tremendous variety of living things on Earth. More than 1.5 million kinds of living things have been described, and more are discovered every year. How can we keep track of them all? Indeed, how can we even find names for them all when their number exceeds the number of words in our language? To the curious, the variety of living things raises many questions.

An initial problem that confronts anyone trying to understand more about life is how to develop a way of describing, sorting, grouping, and naming the various kinds of living things.

Objectives

1. To become more aware of the variety of life that exists on Earth.

2. To understand how living things can be described, classified, and named.

3. To develop a classification scheme for a group of specimens, based on characteristics (observable features, qualities).

Materials

25 numbered specimens
hand lens
compound microscope

Procedure

1. Select a partner to work with and begin at any one of the 25 numbered specimens. **Caution: Use care when handling glass, the specimens, and the microscope.**

2. Plan to spend 2 to 3 minutes observing and listing characteristics of each specimen before proceeding to the next specimen. Use separate sheets of paper to record your observations. You may wish to make a quick line drawing of each specimen, noting major characteristics. Your problem is to identify and list for each specimen characteristics that will eventually be used to sort the specimens into separate groups based on the presence or absence of those characteristics. In other words, the central question you are trying to answer is *How can living things be classified?* This is the question that confronted such naturalists as John Ray, Carolus Linnaeus, and Joachim Jung in the 1600s and 1700s.

3. For each specimen, record whether it is now, or was once, alive.

4. Following your observations, spend about 20 minutes grouping the 25 specimens into six to nine separate groups, each based on a specific set of characteristics. Your major problem here is to decide just which characteristics will be most useful for the classification.

5. List the characteristics used for your classification.

_____

_____

_____

_____

6. Record your classification system and be prepared to report on it in a class discussion.

Study Questions

1. List several characteristics that you could use to determine whether something is alive.

2. Why do people classify things?

3. Give two examples of useful classification schemes used in daily life. Explain why these schemes are important.

4. What might be some reasons for classifying organisms?

_____

_____

_____

5. At what level of the classification hierarchy used by biologists would you expect to find the greatest diversity? Explain.

_____

_____

_____

6. At what level of the classification hierarchy would you expect to find organisms that are most similar? Explain.

_____

_____

_____

7. Two young rats (one male and one female) are captured in the wild and brought to your lab. The biologists who captured the rats have asked you to determine whether the rats are members of the same species. What would you do to find out?

_____

_____

_____

_____

Investigation 3 *How Do Characteristics Vary?*

Introduction

> There is no such thing as an average human.
> Each one of us is a unique individual.
> Each one of us expresses our humanity in
> some distinctly different way. The
> beauty and the bloom of each human soul
> is a thing apart . . . never once repeated
> throughout all the millenniums of time.
> —Lane Weston

When organisms in a particular location share many characteristics, interbreed in nature, and produce fertile offspring, they are said to be all of one kind, or one *species*. But how similar are all the members of a single species? Obviously, they are not all identical. Humans are all one species, yet differences among us in skin color, eye color, height, weight, and so on are obvious. Or, for example, suppose you find two trees that appear the same except one has leaves 13 cm across and the other has leaves 3 cm across. Are the trees still of the same species? In what ways can individuals of a single species differ? Are these differences systematic or random? What causes these differences?

Objectives

1. To investigate and discover patterns of variation within species.

2. To learn to graph data that organize and summarize differences within species.

3. To propose and initially test alternative hypotheses to explain the observed patterns of variation.

Materials

bag of assorted shells
metric rulers
graph paper
pair of dice
ear of Indian corn

Procedure

1. Work with a group of two to three other students. Empty the contents of the bag of shells onto a table.

2. Sort the shells at your table into groups that you believe represent separate species. List characteristics used for your sorting.

_____

_____

_____

_____

3. How many species do you think are present?

4. Select one species that exhibits differences that can be noted and/or measured.

5. Collect as many individuals of that species as you can from throughout the room.

6. For each shell, determine the value of the characteristic chosen. Record the frequency of occurrence of the values (measure at least 50 to 60 individuals—the more the better).

7. What characteristic did you measure?

8. Plot your data on a frequency graph. Title the graph and label both axes. Use a separate sheet of graph paper.

9. What is the average, median, mode, and range of your data? (Note: Your teacher may want you to skip this step until you have completed steps 10 – 15).

10. The graphs for each species will be collected, put on the board, compared, and contrasted. The graphs should enable you to answer the question, *In what ways do characteristics vary?* They should raise additional questions such as, *What causes the observed patterns of variation?* To help answer this last question, do the next activity with the dice.

11. Roll two dice. Note the numbers that turn up on each die. Add these numbers and record this sum. Repeat 100 times and plot the data on a frequency graph. Compare this graph with the previous graphs. Do these data suggest a mechanism for the observed variation within species? Explain.

12. Obtain an ear of Indian corn. Count the number of kernels of each color on the ear. How many kernels of each color were there?

13. Plot these data on a frequency graph. Use a separate sheet of graph paper. Approximately what ratio of colors did you find?

14. Compare your graph with your previous graphs. How are they similar? How are they different?

15. Compare your results with those of other students. Do these data suggest a cause for the observed variation? Explain.

Study Questions

1. What is the difference between a sample and a population?

2. How could you get a representative (unbiased) sample of the population of your school's students?

3. What is a normal distribution? Why is it called normal?

4. Why do characteristics often distribute themselves normally?

5. What are genes? Where are they located?

6. What is meant by dominant and recessive characteristics?

7. Human intelligence is the result of an individual's genetic makeup and his or her environment. Give some reasons why it is difficult to determine the relative contributions of the genes and the environment to human intelligence.

8. *Genotype* refers to the collection of genes in an individual's body cells. Each body cell has the same collection of genes—the same genotype. Sex cells, or *gametes,* have only half of the genes of a body cell. Look at the following genotypes. How many different kinds of gametes are possible? Assume that in this case the genes assort themselves independently of one another (i.e., no linkage). Dominant genes are represented by uppercase letters, recessive with lowercase.

   Parental Genotypes:

   a. Dd   **b.** DdEe   **c.** DdEeff

   _____

   _____

   _____

   _____

9. What is the probability of getting the gamete de from each of the following genotypes?

   a. DDEe   **b.** Ddee   **c.** ddee

   _____

   _____

   _____

   _____

10. What is the probability of each of the following sets of parents producing the given genotypes in their offspring?

Parents		Offspring Genotype

    a. DD x Dd   Dd _____

    b. Dd x Dd   Dd _____

    c. DdEe x DdEE   DdEe _____

    d. DdEe x DdEe   DDEE _____

11. In cats, the gene for lack of color is recessive to the gene for normal coloration. Suppose two *heterozygous* (one of each kind of gene) cats have kittens.

    a. A kitten has normal coloration. What is the probability that it is a carrier of the recessive gene?

    _____

    _____

    b. What is the probability that a kitten will lack color?

    _____

    _____

12. In certain bean plants, the gene L for large pods is dominant over the gene l for short pods.

    a. If both individuals are heterozygous (L l), what will be the genotypic ratios of the offspring? What about the phenotypes — the characteristics you can see? What will be the phenotypic ratios?

_____

_____

    b. If a *homozygous* (both genes are the same) long bean plant (L L) and a homozygous short bean plant (l l) are crossed, what will be the genotype and phenotype of the offspring?

_____

_____

13. Eye color of the imaginary grizzly gronk population varies. Some gronks have purple eyes, some have white eyes, and some have orange eyes. Professor Greengenes has discovered that whenever two purple-eyed gronks mate, they always produce purple-eyed offspring. Likewise, whenever two orange-eyed gronks mate, they always produce orange-eyed gronks. But when white-eyed gronks mate, they are able to produce offspring with three colors of eyes.

    a. Use Mendelian theory to explain how eye color is determined among gronks; that is, what is the genotype of the white-eyed gronks, and how can they mate to reproduce offspring of all three colors of eyes?

_____

_____

_____

_____

_____

    b. Use your theory to predict the phenotypic ratio of offspring if purple-eyed and white-eyed gronks were mated.

_____

_____

_____

_____

Reasoning Module 1 *Probability*

Introduction

What are your chances of winning a $5,000,000 sweepstakes? What are your chances of being struck by an automobile while crossing the street? What are your chances of living to be 100 years old? Do frogs with spots have a better chance of survival than frogs with no spots? Does a farmer have a better chance of raising a good corn crop if he or she sprays the crop with insecticide? The notion of probability is fundamentally embedded in every part of our lives, as well as that of every other living organism. Understanding probability is therefore fundamental to understanding life and the world in which you live.

Objective

To develop an understanding of probability and its importance to an adequate understanding of the world.

Materials

sack of brown and white beans

Procedure

1. Begin by reading the first essay and answering the questions.

2. Continue by reading the second essay and doing the activity provided.

3. Work through as many study problems as you think necessary to obtain a good understanding of the ideas introduced.

Essay 1: The Tennis Balls

Imagine a girl walking down a sidewalk and bouncing two tennis balls—a white one and a yellow one. Each time she bounces the two balls, the yellow one bounces higher than the white one even though she drops them both from the same height.

As you watch, a second girl, carrying a tennis racket, runs up to the first girl and says, "Hey, Sis, give me back my yellow tennis ball. It's time for my tennis lesson."

At hearing this, the first girl replies, "No. I want it. You can have the white tennis ball."

The second girl exclaims, "I don't want the white ball. It doesn't bounce worth beans."

To this, the first girl says, "I'll tell you what. I'll hold both tennis balls behind my back—one in each hand. If you can guess which hand has the yellow ball, you can have it."

The second girl replies, "OK, but hurry up. I'm late. I'll pick the left hand."

Question: What are the second girl's chances of correctly guessing the hand with the yellow ball?

The answer is that the chances are 1:1, 50-50, 50%, or one-half, depending on how you say it.

In this situation, the second girl can make two possible choices—the right hand or the left hand. These choices are called *possible events*. Picking the right hand would be one possible event. Picking the left hand would be another possible event.

Picking the yellow ball is the event the second girl wants. This event is one of the two possible events, so the second girl's chances (probability) of getting what she wants is 1 out of 2, which can be written as 1/2.

$$\frac{1 \text{ (event of picking yellow)}}{1 \text{ (event of picking yellow)} + 1 \text{ (event of picking white)}} = \frac{1}{2}$$

The other answers, such as 50%, 1:1, or 50-50, also describe the chances (the probability). Fifty percent means that if you guessed 100 times, you would guess correctly about one-half of the time, or 50 times (50 per 100 = 50%). One to one means for every one event that is correct (picking the hand with the yellow ball), there is one event that is not correct (picking the hand with the white ball). Fifty-fifty means if you guessed 100 times, one-half of those times, or 50 times, you would pick the yellow ball, and one-half of the time, or 50 times, you would pick the white ball.

Essay 2: A Sack of Beans

As you learned in the previous essay, the probability (chances) of the second girl picking the yellow tennis ball was 1 out of 2, or 50%. But what does this 50% mean anyway? Does it mean that if the second girl actually guessed 100 times which hand held the yellow ball she would guess correctly exactly 50 times? Let's see just what this means through an example activity.

Pick up the sack of beans provided for this activity. Look inside but do not count the beans. Notice the two colors of beans—white and brown.

Now close the sack and shake it up. Without looking into the sack, pick out ten beans at random. How many of the beans are brown? Plot this number on a bar graph like the one shown below. Now put the ten beans back into the sack and shake the sack again.

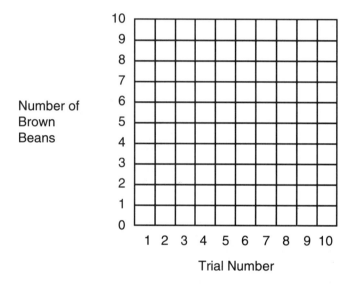

Pick out another ten beans at random. How many of these are brown? Plot this number on the graph. Again, put the ten beans back into the sack and shake the sack.

Repeat this procedure for eight more trials.

Now look at your graph.

Does it appear as though "on the average" 5 out of every 10 beans are brown?

Is this the same as saying "on the average" 1 out of every 2 beans is brown?

Does 5/10 = 1/2?

Suppose you are told that the sack contained exactly 100 beans. How many of the beans will you predict are brown?

Count the beans and see how accurate your prediction is. If your prediction is not exact, how can you account for the inexactness?

In the above activity, the 100 beans can be called a *population*. You obtained ten different samples of the population. For each sample, it is possible to estimate (guess) characteristics of the population. In this case, you attempted to estimate the portion of the population that was brown. That portion was about 50 out of 100, or 5 out of 10, or 1 out of 2. Therefore, in this population, the probability of being brown is about 50%. Notice, however, that this is just an estimate. The sack did not contain exactly 50 brown beans. All estimates contain some error. The idea is to make the error as small as possible by getting as large a sample as possible without going to too much trouble or expense.

The probability of any desired event(s) is given as follows:

$$\text{Probability} = \frac{\text{Number of desired events}}{\text{Total number of possible events}}$$

Multiplying the result by 100 gives the probability expressed as a percentage.

Study Questions

1. If you flip a coin, what are the chances it will come up heads?

_____

Again, the probability is 1/2, or 50/100, or 50%. The event "coming up heads" is what we want. Two equally possible events exist, heads and tails, so the answer is one event out of two possible events.

2. When you roll a die, what is the probability you will roll a 6?

_____

The event you want is "rolling a 6." What are the possible events? Rolling a 1, rolling a 2, rolling a 3, rolling a 4, rolling a 5, rolling a 6—a total of six different possible events. The chances of rolling a 6, then, are 1 out of 6, or 16.7% (1/6 = 1 ÷ 6 = .167; .167 x 100 = 16.7%).

3. Suppose 5 real diamonds and 20 fake diamonds are put into a sack. What are your chances of reaching into the sack and pulling out a real diamond on the first try?

_____

_____

The event you want is "picking a real diamond"? This can be done in five ways because you could pick any one of five real diamonds. Twenty-five events are possible because 25 pieces in all are in the sack, so your chances of "picking a real diamond" are 5 out of 25, or 20% (5/25 = 20/100 = 20%, or 5/25 = 5 ÷ 25 = 20%).

4. Suppose you roll two dice. What are your chances of rolling a 12? Of rolling a 7?

_____

_____

5. A farmer collects 25 field mice from one corner of a field and notes that 20 of the mice are spotted and 5 are all-white.
   a. How many mice are in the farmer's sample? What fraction of the sample of mice is spotted?

_____

_____

b. What percentage of the sample is spotted?

_____

_____

c. Suppose the farmer captures one more mouse. What is the probability that the mouse will be spotted?

_____

_____

6. In a small town in Arizona, 75 people bought a lottery ticket from the local grocery store. Three of those 75 people bought winning tickets.
   a. What portion of the people bought winning tickets?

_____

   b. Suppose you were to go to that grocery store and buy a lottery ticket. What would you estimate your chances are that your ticket is a winner?

_____

7. On the first day of a college biology class for nurses, 23 out of the 28 students enrolled were females.
   a. What percentage of the class was female?

_____

_____

   b. On the second day of class, a new student enrolled. What are the chances that this student was a male?

_____

_____

   c. Does this mean that only about 18% of the population of students at this college are male? Explain your answer.

_____

_____

Brain Bender

What are your chances of finding a needle in a 3-meter-tall haystack within 1 hour? What variables would you need to consider? What assumption could you make before generating a probability?

_____

_____

Investigation 4 *Which Characteristics Are Linked?*

Introduction

Take a look around the classroom at the other students. In what ways do their characteristics vary? Some are tall, some are short, and some are in-between. Height is a characteristic that varies. How about eye color? Some eyes are brown, some are blue, some are green, and some are in-between. Eye color varies. What other characteristics can you name that vary? Can you name some that do not vary?

In this investigation, you will measure and record a number of human characteristics that vary, such as height, "eyedness," handedness, and quickness, and attempt to discover whether any of the values of the variables seem to vary together (are "linked"). Much of biology is concerned with the search for links between variables because they represent clues to help us better understand nature. Does a link exist between eating foods high in cholesterol and your chances of having a heart attack? If so, perhaps cholesterol somehow acts to cause heart attacks and should be avoided in your diet.

Learning about how biologists search for links between variables and possible cause-effect relationships is the major purpose of this investigation.

Objectives

1. To learn about an important aspect of biological investigation—the search for and establishment of links and possible cause-effect relationships between variables.

2. To analyze and interpret data from two variables in graphs and tables.

Materials

paper with zeros
timer
3-by-5-inch card with a hole
meterstick
weight scale (metric if available)

Procedure

1. A number will be assigned to each student in the class. Record your number. You will now gather, record, and analyze data on several characteristics to try to answer the question, *Which characteristics are linked?*

2. **Gender** — Record your gender (M or F) on a data table that lists the variable names at the top and the student numbers along the left side and on the class data table.

Variable Names

Student	Gender (M or F)	Handedness (L or R)	Eyedness (L or R)	Quickness (L or R)	Height (cm)	Weight (kg)	Arm Span (cm)	Rolling (Yes or No)
1								
2								
3								
4								
5								
6								
7								
8								
9								
10								
11								
12								
13								
14								
15								
16								
17								
18								
19								
20								
21								
22								
23								
24								
25								
26								
27								
28								
29								
30								
31								
32								
33								
34								
35								

3. **Handedness**—Obtain a photocopy of the rows of zeros shown on the next page. Have your partner use slash marks to cross out, one at a time, as many zeros as he or she can in 30 seconds, using the right hand. Repeat, using the left hand. Record those numbers. Calculate the "handedness measure" for each person. (Handedness measure is defined as the number of zeros crossed out by the higher scoring hand divided by the number crossed out by the other hand. Designate whether the right or left hand was the higher-scoring hand.) Record your calculation on your data table and on the class data table. Note: Each line has 62 zeros.

OOO

OOO

OOO

OOO

OOO

OOO

OOO

OOO

4. **Eyedness** — Hold the 3-by-5-inch card at arm's length. With both eyes open, look through the hole at some distant object. Close your left eye. Can you still see the object? If so, you are right-eyed. Open your left eye and close your right eye. Can you now see the object? If so, you are left-eyed. Record your finding.

5. **Quickness** — Obtain a meterstick and two pencils. Have your partner hold the meterstick as shown. Your finger and thumb should be positioned at the 20-cm mark, far enough away from the stick to allow a pencil to pass on either side of the stick. This should be the starting position for each trial. Your partner should then drop the stick with no warning. When he or she does, catch the stick without moving your arm. Practice this a few times, until you get a feel for how it is done. Try to catch the stick as quickly as possible. The number of centimeters from the 20-cm mark to where you catch the stick is a measure of your quickness. Measure quickness six times for each person, for each hand. Obtain a mean quickness measure for each. Record.

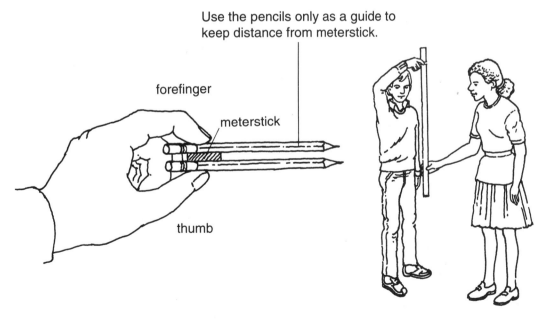

Use the pencils only as a guide to keep distance from meterstick.

forefinger

meterstick

thumb

Place your fingers at the 20-cm mark, ready to catch the stick.

Hold the stick at the 100-cm end.

6. **Height** — Measure your height in centimeters. Record. You may omit this measurement if you choose.

7. **Weight** — Measure your weight in kilograms or pounds. Record. You may omit this measurement if you choose.

8. **Arm Span** — While standing, extend both arms straight out at your sides and parallel to the floor. Measure the distance from fingertips on the right hand to fingertips on the left hand (in centimeters). Record.

9. **Tongue Rolling** — Can you curl the sides of your tongue? If you can, you are a tongue roller. If you cannot, you are not a tongue roller.

10. Record the data for all class members on your data sheet.

11. Select from the data two *continuous variables* (variables with numerous values probably determined by multiple gene pairs) that you think might be related. Why do you think they might be related?

_____

_____

12. **a.** Using the method demonstrated by your teacher, plot a graph of the two variables.
    b. Does the graph show a relationship between the variables? Explain.

_____

_____

   c. If so, is the relationship direct or inverse? Explain.

_____

_____

13. Select from the data two dichotomous variables (variables with only two values probably determined by a single pair of genes) that you think might be related. Why do you think they might be related?

_____

_____

14. Using the method demonstrated by your teacher, construct a contingency table of the two variables. Does the table show a relationship between the variables? Explain.

_____

_____

Study Questions

1. What is a variable? How is it similar to or different from a characteristic? A trait? A property?

_____

_____

2. Name four variables and at least two values for each.

_____

_____

3. What is a constant?

_____

_____

4. Which of the variables measured in this activity are continuous? Discontinuous? Both, depending upon the measurement?

_____

_____

_____

_____

5. Which of the variables are probably determined by a single gene pair? By many gene pairs? Explain.

_____

_____

_____

_____

6. A biologist was observing the mealworms that live in a grain elevator near Omaha. She noticed that some of the mealworms were large, while others were fairly small. Also, some of the mealworms had spots, while others appeared to have no spots.

She wondered whether a correlation might exist between the size of the mealworms and the number of spots on their backs, so she collected a sample of mealworms to see. Below are pictures of the worms she collected.

   a. Construct a contingency table of the mealworm data on a separate sheet of paper.

   b. Do the data indicate that a correlation exists between size and number of spots? Explain.

_____

_____

_____

7. Read the following newspaper story.

    a. Name the variables that are the primary focus of the article. What relationships have been found? How do the researchers try to explain their results?

    b. Obtain a list of names of the members of your school's baseball and/or softball teams along with their batting averages for the previous season. With the assistance of classmates, test each team member for eye and hand dominance. Share the results of the tests with others in the class. Analyze the results to find out whether or not results similar to those described in the article are found. You may want to report your analysis to the players and coaches.

Keep your eye on the ball, sure, but which eye?

Students of game focus on pupils

The Associated Press

BOSTON—Baseball sluggers see the world slightly differently from the rest of us, and that helps explain why they can hit the ball so well, according to a new study.

The report, published as a letter in today's *New England Journal of Medicine,* attributes ballplayers' slugging skills to "crossed eye-hand dominance."

Just as people prefer to use one hand over the other, they also tend to see from one eye more than the other. If people are both right-handed and right-eyed, they are said to have uncrossed eye-hand dominance. If the favored hand and eye aren't on the same side, they have crossed dominance.

Drs. Jose M. Portal and Paul E. Romano of the University of Florida College of Medicine gave eye tests to ordinary people and found that two-thirds had uncrossed eye hand dominance. The other third were evenly split between those with crossed dominance and those who used both eyes equally.

However, they found that the university's baseball team members were twice as likely to have crossed dominance and 50 percent more likely to use their eyes equally.

"We think crossed dominance may aid the batter," the doctors wrote.

They found that players with crossed dominance had batting averages of .310 compared with .250 for those with uncrossed dominance. However, those who saw equally from both eyes had the best average, .340.

While players with uncrossed dominance are worse hitters, they also tend to be better pitchers.

"Uncrossed dominance may therefore be a relative handicap to hitting, forcing those who have it to become pitchers, or it may be an aid to pitching," they wrote.

The researchers cautioned that people should not try to change their dominant eye, because this can cause permanent double vision and other eye problems. And vision dominance may be inherited, anyway.

However, they said that if confirmed, their findings might be used to guide young athletes into the sports specialties they are best suited for.

"More specifically," they wrote, "since batters with crossed dominance seem most successful, they could try a crossed-dominant stance: If right-eyed, bat left-handed, and if left-eyed, bat right-handed."

Reprinted with permission of the Associated Press.

Reasoning Module 2 *Correlation*

Introduction

Does smoking cause lung cancer? Does laetrile cure cancer? Does drunk driving cause accidents? Do fatty foods cause high blood pressure? Does increased exposure to radiation increase mutations? Do thin people live longer than heavy people? Do small organisms have higher metabolic rates than large ones?

You may think you know the answers to some of these questions already. On what evidence and/or reasoning do you base your answers? The collection and analysis of data to determine whether two variables are "linked" are important components of scientific investigation. Finding variables to be linked (correlated) suggests the possibility of cause-effect relationships, which are at the very heart of human understanding.

Objective

To learn how to analyze data for correlations and possible cause-effect relationships.

Materials

packets A through F

Procedure

1. Begin by reading the first essay. Do the suggested activities and answer the questions.

2. Work through the other essays and some or all of the study questions until you are satisfied you understand the terms introduced.

3. For assistance, consult your teacher or a student near you.

Essay 1: Things That Go Together

Recall the girl walking down the sidewalk and bouncing the white and the yellow tennis balls. Her sister finally did get the yellow ball and left for her tennis lesson. As you continue to watch, the girl starts walking down the sidewalk until she comes to a bumpy dirt road. She keeps walking down the road and starts to bounce the white tennis ball as she goes along. You notice that each time she drops the ball, it hits a different part of the road. Sometimes the ball hits a soft spot and a low bounce results.

Because the height the ball bounces changes (varies) from bounce to bounce, the "height of bounce" is a called a *variable*. If the height of the bounce stayed the same, it would not be called a variable. It would be called a *constant*.

Notice that the condition of the road also varies. Sometimes it is hard, and sometimes it is soft. So "hardness of the road" would also be considered a variable. Of course, if the road were paved and so was equally hard in all spots, the "hardness of the road" would not be considered a *variable*. It would be a *constant*.

In the present example are two variables: 1) the distance the ball bounces after it hits the road, and 2) the condition of the road. The ball bounces either high or low and the road is either hard or soft.

Since the two variables "go together" (high bounces occur when the ball hits the hard spots, and low bounces occur when the ball hits the soft spots), a correlation exists between the variables "height of bounce" and "hardness of road."

A brief statement summarizes the "linked" or correlational relationship: The harder the road the higher the bounce or, said another way, the softer the road the lower the bounce.

Now that you have some idea of what is meant by the terms *variable* (something that changes), *constant* (something that stays the same), and *correlation* (two things that vary together), let's see how these ideas can be applied in a new situation.

Essay 2: The Strange School in Lakeville

Imagine you are visiting a friend who is a student at Lakeville High, a somewhat strange school located in Lakeville. I say strange because the students at Lakeville High never choose the classes they take. They are assigned to classes based on their hair color and eye color. Also, every course has exactly 100 students enrolled. Because this all seems so strange, you decide to investigate some of the classes.

English — The first class you visit is English. Pick up Packet A, which contains pictures of a sample of 20 of the 100 English students. Sort the sample of students into two groups based on observed differences.

- What characteristic(s) did you use for your sorting?

- How many students are in each group?

Because hair color varies among the English students (sometimes it is blond, sometimes it is brunet), hair color is considered a variable. English students have either blond or brunet hair. Eye color, however, is constant; it is always brown. Apparently, you need brown eyes to enroll in English.

If our sample of 20 students is unbiased, we can extrapolate the trend in the data to include all 100 students in English. If we do this, about 50 out of 100 (50%) have blond hair and about 50 out of 100 (50%) have brunet hair; 100 out of 100 (100%) have brown eyes.

Biology — The next class you visit is Biology. Pick up Packet B and sort the pictures of 20 Biology students.

- What characteristic(s) did you use for your sorting?

- How many students are in each group?

Because eye color varies in the sample of Biology students (sometimes it is blue, sometimes it is brown), eye color is a variable. Notice that hair color is constant; it is always brown. Apparently, in Biology, you can have either blue or brown eyes, but you must have brown hair.

Again, if we extrapolate the data from 8 out of 20 to 40 out of 100, we can say that about 40% of the students in Biology have blue eyes. Twelve out of 20 or 60 out of 100 (60%) have brown eyes.

Physical Education — Pictures of a sample of Physical Education students are in Packet C. Sort these students into two groups based on observed differences.

- What characteristic(s) did you use for your sorting?

- How many students are in each group?

Notice that hair color and eye color are both variables in this class. Also notice that they go together (covary). The blond-haired students all have blue eyes, and the brunet-haired students all have brown eyes. Ten out of 20 or 50 out of 100 (50%) of the students are blue-eyed blonds, and 10 out of 20 or 50 out of 100 (50%) are brown-eyed brunets.

We say that a correlation exists between "hair color" and "eye color" in this sample, just as a correlation existed between the variables "height of bounce" and "hardness of the road" in the first essay.

Apparently, to be enrolled in Physical Education, you must have either blond hair and blue eyes or brunet hair and brown eyes.

History — Let's try another class and see what other relationships we can find. Pick up Packet D, which contains pictures of a sample of 20 students from History class. Sort these students into two groups.

• Do hair color and eye color go together?

_____

Again, the answer is yes although not in the same way as before. Now, all blond-haired students have brown eyes, and all the brunet-haired students have blue eyes. Nevertheless, hair color and eye color do go together in the History students.

• What percentage of the total is in each group?

_____

Apparently, to be in History class, you must have either blond hair and brown eyes or brunet hair and blue eyes.

Geometry — A sample of 20 pictures of the students from Geometry are in Packet E. Sort these students into groups.

• How many different groups were there? How many students are in each group?

_____

• Do hair color and eye color go together in this sample?

_____

The answer is yes. Hair color and eye color do go together, but two students did not fit the pattern—the student with blond hair and brown eyes and the student with brunet hair and blue eyes. These students are exceptions. Nevertheless, because most of the students fit the rule (18 out of 20 or, if we extrapolate the trend, 90 out of 100 [90%]), it still looks as though hair color and eye color go together (are correlated) in Geometry. Eighteen students fit the rule, and only two are exceptions.

Said another way, most (9 out of 10, or 90%) of the blond-haired students have blue eyes, while most (9 out of 10, or 90%) of the brunet-haired students have brown eyes, so the variables go together — most of the time, anyway.

Said still another way, the probability of having blue eyes if you have blond hair is 90%. The probability of having brown eyes if you have brunet hair is also 90%.

Just why we have two exceptions to the rule is not clear. Maybe the two students sneaked into class without the teacher noticing.

Music — Pick up Packet F with pictures of students from Music in it. Sort the students into groups.

• How many groups are there? How many students are in each group?

_____

• Do hair color and eye color go together (covary) in Music?

_____

In this instance, the answer is no. Five students are in each of the four groups. We have just as many students who are exceptions (10) as students who fit the pattern (10). In at least two ways we can explain why the data indicate that hair color and eye color do not go together.

1. Suppose you thought that blond hair and blue eyes should be one group and brunet hair and brown eyes should be another group. Ten students fit this expectation, but ten students do not (the blond hair/brown eyes and the brunet hair/blue eyes). Because 10 out of 20 (50%) fit and 10 out of 20 (50%) do not fit, no correlation exists.

2. Can we say that most of the blond-haired students have blue eyes and most of the brunet-haired students have brown eyes? No, this is not the case. Only 10 out of 20 (50%) of the blond-haired students have blue eyes, and only 10 out of 20 (50%) of the brunet-haired students have brown eyes. Again, it is 50% to 50%, so it looks as though no correlation exists. The probability of having blue eyes is 50% whether you have blond hair or brunet hair. The probability of having brown eyes is also 50% whether you have blond hair or brunet hair. Hair color and eye color do not seem to go together in this class.

Let's work through one more problem. Then you may try a few others if you still feel somewhat unsure of some of the ideas presented.

Essay 3: The Swim Team Tryouts

Last spring, 25 girls showed up to try out for the girls' swim team. To qualify, each swimmer was to swim the 50-m freestyle race, with the 15 fastest swimmers making the team.

The coach noticed that often the taller girls would beat the shorter girls in the qualifying races. This made her wonder whether a correlation might exist between height of the swimmers and the time it took them to swim the 50 meters. To find out, she recorded the swimmers' heights and their respective times for the 50-m race, as shown in the table below.

Then the coach plotted the data on a graph, as shown in Figure 1.

Swim Team Tryouts

Swimmer	Height (feet & inches)	Time (in seconds)
1	5'2"	28.0
2	5'0"	29.5
3	5'0"	30.5
4	5'7"	26.5
5	5'10"	26.5
6	5'3"	25.0
7	5'8"	25.0
8	5'11"	24.0
9	6'0"	24.5
10	5'3"	29.9
11	5'1"	26.0
12	5'6"	27.0
13	4'11"	31.0
14	5'10"	28.5
15	5'11"	25.5
16	6'1"	25.5
17	6'2"	24.0
18	5'4"	30.5
19	5'2"	29.5
20	5'5"	29.0
21	5'7"	28.0
22	5'8"	30.5
23	6'0"	26.5
24	5'5"	28.0
25	5'0"	28.5

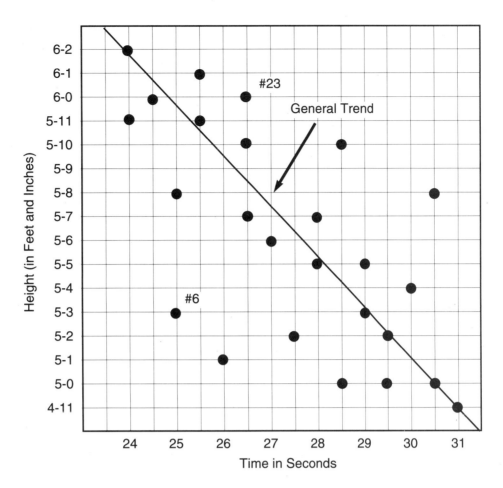

Figure 1. Swimmers' heights versus time to swim 50 meters.

The graph shows a general inverse correlation between the variables "height" and "time." Although exceptions occur (e.g., swimmer #6 is shorter yet faster than swimmer #23), in general, the taller the swimmer, the less the time it took to swim the race.

The coach then divided the graph into four sections as shown in Figure 2. The groups are

    1. the swimmers who were tall and fast,

    2. those who were tall and slow,

    3. those who were short and fast, and

    4. those who were short and slow.

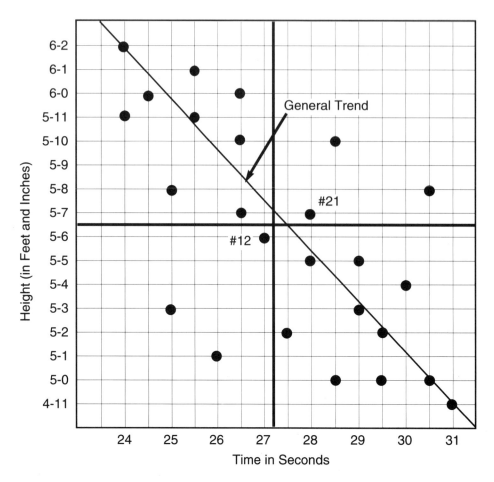

Figure 2. Correlation between height and time.

If an inverse correlation exists, we would expect to find most swimmers in the tall and fast or short and slow categories.

- Is this the case?

In general, most of the tall swimmers (9 out of 12) were fast and most of the short swimmers (10 out or 13) were slow. Said another way, 19 of the swimmers were either tall and fast or short and slow, while only 6 swimmers were exceptions (tall and slow or short and fast). So, the variables "height" and "time" do go together. A correlation exists between them.

- Do you think it is fair to categorize swimmers #12 and #21 as exceptions? Explain.

_____

_____

- Does height cause a swimmer to be fast?

_____

_____

The issue of determining cause-and-effect relationships will be discussed later.

Study Questions

1. On the next page is a graph that shows the speed at which automobiles were driven and the number of miles obtained from 1 gallon of gasoline.

a. Does the graph show a correlation? Explain your answer.

_____

_____

_____

b. Are any points exceptions? If so, what are they?

_____

_____

c. What could be the reason for the exceptions?

_____

_____

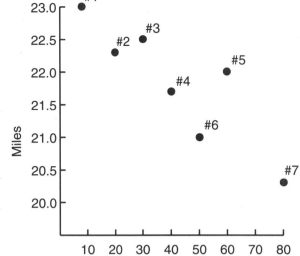

2. Below is a sample of frogs collected from a pond. Does the sample show a correlation between the size of the frogs and the number of spots on their backs? Explain your answer.

_____

_____

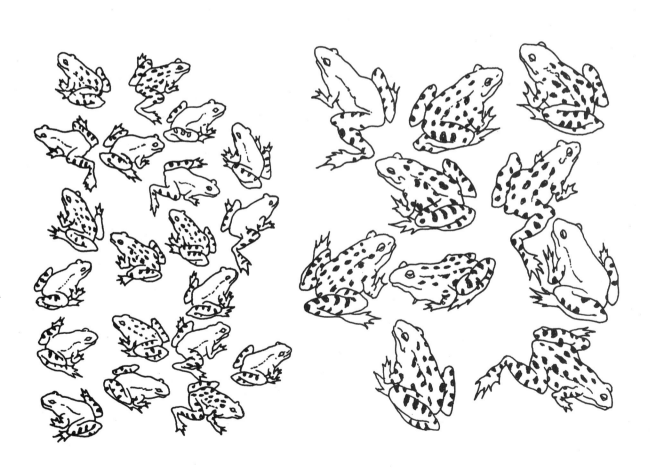

3. The medical records from County General Hospital revealed the following data. The patient numbers were selected randomly from all of those who died at the hospital in the past 10 years.

Patient Number	Cause of Death	Smoker Yes	Smoker No
45	Lung Cancer	X	
51	Internal Bleeding		X
73	Lung Cancer	X	
84	Natural Causes		X
92	Lung Cancer	X	
93	Lung Cancer		X
103	Heart Attack		X
104	Lung Cancer	X	
121	Natural Causes		X
132	Heart Attack	X	
143	Stab Wounds	X	
150	Lung Cancer	X	
157	Breast Cancer		X

Construct a contingency table of the data. Do the data show a correlation between death by lung cancer and smoking? Explain your answer.

_____

_____

Aspirin tied to reducing heart woes

Effects dramatic in men, study says

By HAROLD M SCHECK Jr.

The New York Times

NEW YORK — A major nation-wide study strongly suggests that a single aspirin tablet every other day can sharply reduce a man's risk of and death by heart attack.

The lifesaving effects were so dramatic that the study was halted in mid-December so the results could be reported as soon as possible to the participants and to the medical profession in general.

The magnitude of the beneficial effect was far greater than expected, said Dr. Charles H. Hennekens of Harvard Medical School, principal investigator in the research. The risk of myocardial infarction, or heart attack, was cut almost in half.

A special report said the result showed "a statistically extreme beneficial effect" from the use of aspirin.

The report is to be published Thursday in *The New England Journal of Medicine*. The findings were not scheduled to be made public until 6 p.m. today, but details of the report were disclosed Tuesday by the Reuters news agency, which said it had learned about the results from unidentified industry sources.

In recent years, smaller studies have demonstrated that a person who has had one heart attack can reduce the risk of a second by taking aspirin, but there had been no proof that the beneficial effect would extend to the general male population.

Dr. Claude Lenfant, director of the National Heart, Lung and Blood Institute in Bethesda, Md., said the findings were "extremely important," but he said the general public should not take the report as an indication that everyone should start taking aspirin.

"It should be used as a drug, not a panacea," he said.

He and other specialist stressed that the decision on using aspirin to prevent heart attack should be made on an individual basis by the person and his physician.

The study was by far the largest ever done to assess the effects of aspirin on heart disease, and the first to gauge the benefit for healthy people, Hennekens said.

Participants were 22,071 physicians recruited through lists made available by the American Medical Association. Their state of health and health consciousness was generally excellent. In this extremely low-risk group, the benefits of aspirin stood out with unexpected vividness.

Even before this study began, many people believed that taking aspirin would be beneficial. Hennekens said about half the physicians eligible for the study were ruled out because they were already taking aspirin.

Participants took either one buffered-aspirin tablet every other day or a placebo. Those in the study did not know which of the two they were receiving.

A panel of the medical scientists who were doing the research checked the results every six months. The latest such review, in December, convinced the committee that the study had proved its main point concerning aspirin and that aspect of the study should be halted so those who were receiving the placebo could switch to aspirin if they wished.

More than 22,000 physicians were involved; 11,037 received aspirin, and 11,034 took the placebo.

Among those who received aspirin, there were five fatal heart attacks and 99 non-fatal attacks. In the "placebo group," there were 18 heart-attack deaths and 171 non-fatal attacks.

In its relationship to heart attacks, aspirin is believed to act by inhibiting the aggregation of platelets, blood particles that are important in the process of blood clotting. The presumption is that the protective effect is a result of the reduction in clot formation.

This same mechanism also is a potential hazard because it could decrease blood clotting too much in some individuals. Indeed, the study showed that there was a small, although not statistically significant, excess of death from strokes among those who had received aspirin.

The presumption was that these were strokes caused by hemorrhage in the brain. Among those who received aspirin, there were six fatal strokes, as against two among those who received placebos.

Scientists said they believed the excess of strokes was real even though it was not sufficiently strong to be statistically significant. The finding of a small excess of strokes was one of the reasons Hennekens and Lenfant both said the decision for or against using aspirin to prevent heart attack should be made on a case-by-case basis.

4. **a.** What two variables are presumed to be correlated in this article?

   b. Set up a contingency table for the two variables. Analyze the data to determine whether they show the variables to be correlated.

   c. Does the author also argue that one variable is a cause? If so, what mechanism is suggested (how does it act to cause the effect)?

   d. Is evidence given that other variables are correlated with taking aspirin? Explain.

Doing poorly

Study links low income with high blood pressure

United Press International

ATLANTA — The less money you earn, the more likely you are to have high blood pressure, a medical researcher says.

Dr. Neil Shulman, an associate professor of medicine at Emory University, says that conclusion is borne out in a study of 72,000 people in Georgia who suffer from moderate to severe hypertension.

"The lower the income, the higher the blood pressure," said Shulman, a co-author of the study. "This correlation is so compelling that it is almost possible to take someone's blood pressure and then figure out how much that person makes a year."

Shulman said the statistics are significant because it means that "those who have the hardest time affording treatment are in a deadly bind."

"They are more likely to have severe hypertension, and they have to spend a much larger percentage of their income on medicine," he said.

Cost, according to Shulman, has become the chief obstacle to treatment for many lower-income Americans.

He said one-third of the people with untreated severe hypertension will develop complications within 11 months. But he said that when he and his colleagues surveyed 72,000 Georgians with moderate to severe hypertension "we found that only 20,000 of them are on medication."

"That's 52,000 people not on medication — a significant number of which will develop severe complications within the next year," he said.

The survey, results of which were published in the *American Journal of Public Health* revealed that blacks are five to seven times more likely than whites to suffer moderate to severe hypertension, with black hypertensives 10 times more likely than white hypertensives to fall prey to end-stage renal disease — kidney failure requiring dialysis and a common complication of uncontrolled high blood pressure.

Reprinted with the permission of United Press International, Inc.

5. a. What two variables are correlated, according to the article?

_____

b. Is this a direct or an inverse correlation? Explain.

_____

_____

c. Does this author argue that one of the variables is a cause? Explain.

_____

_____

6. Look through the newspaper during the next few days. Read and cut out an article that discusses variables that are correlated. Bring the article to class for discussion.

Investigation 5 *What Can Be Learned From Skulls?*

Introduction

Do we need to see an entire animal to determine where it lives or what it eats? Sometimes we can use bones as clues to provide insight into possible answers to these questions. Observation is a key to understanding. What can be inferred by looking at skulls?

Objectives

1. To infer function and animal behavior from observation of skull characteristics.

2. To support or refute alternative hypotheses through use of evidence and logical argumentation.

Materials

10 skulls of 10 different species of vertebrates

Procedure

1. With your partner, go to a work station and take about 5 minutes to examine the skull carefully.
   a. Observe the size and shape of the skull, as well as characteristics of the teeth, eye sockets, braincase, and so on. Record interesting observations on a data sheet. Make a sketch if you want.
   b. For each skull, try to answer these questions: *What kind of animal is this? What did it eat? Where did it live?*
   c. What arguments and evidence do you have to support for your hypothesis (in step b)?

   Station 1 _____

   _____

   _____

2. Move to the next station when you are ready. (No more than two groups should work at one station simultaneously.)

3. Be prepared to discuss your hypotheses, arguments, and evidence in a class discussion.

   Station 2 _____

   _____

   _____

   Station 3 _____

   _____

   _____

   Station 4 _____

   _____

   _____

Station 5 _____

_____

_____

Station 6 _____

_____

_____

Station 7 _____

_____

_____

Station 8 _____

_____

_____

Station 9 _____

_____

_____

Station 10 _____

_____

_____

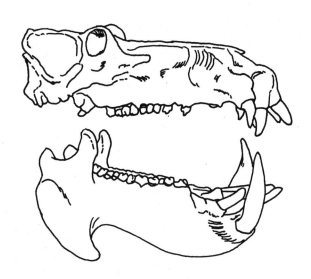

Study Questions

Carefully observe the illustration on the previous page and reflect on your previous laboratory work to answer the following questions.

1. What do you believe the function of the back teeth to be? Why?

_____

_____

_____

2. What do you believe the function of the front teeth to be? Why?

_____

_____

_____

3. Explain why you think this animal is either a herbivore, a carnivore, or an omnivore.

_____

_____

_____

4. Note the position of the eye sockets. How might this positioning be useful to this animal?

_____

_____

_____

5. Can you think of any animals that have this type of eye positioning?

_____

_____

_____

6. Do you believe this animal to be terrestrial, aquatic, amphibious, or adapted for flight? Why?

_____

_____

_____

7. The actual skull is about eight times larger than it appears in the illustration. What animal do you think it is? Why?

_____

_____

_____

Investigation 6 *What Causes Changes in Heart Rate?*

Introduction

When you go to the doctor, he or she may listen to your heartbeat. One of the things the doctor may be interested in is how fast your heart is beating. Why is this important to know? Does your heart rate vary? If so, what are some variables that might influence that rate? Can your body regulate heart rate? If so, how?

Objectives

1. To discover some variables that influence heart rate.

2. To propose a mechanism by which the body regulates heart rate.

Materials

paper bags
step blocks or stairs
clock with second hand

hot water or heating pads
cold water (ice)
buckets

Procedure

1. Determine your pulse rate in beats per minute. Use two fingers, but not your not thumb, to measure at your neck or wrist. Record that number.

2. Use materials listed or additional materials to see how many different things you can do to change your heart rate. For example: *Does breathing into a bag make a difference? Does exercise make a difference?* **Caution: If you have a health condition which may prevent you from doing these activities, be sure to alert your teacher.**

3. Construct a data sheet to record your experiments. The data sheet should include a) the specific question asked, b) what you did, c) the expected result, and d) the actual result. Whenever possible, graph your data and any conclusions that you can draw.

4. How long does it take for your heart rate to return to normal after each experiment?

5. For the procedures that altered heart rate, attempt to create alternative explanations for their effect(s) (i.e., *Why does _____ speed up or slow down heart rate?*). Try to imagine a biochemical mechanism to account for the heart's response to what you did. Be prepared to share your ideas in a class discussion.

Study Questions

1. Propose an explanation for why hyperventilation can be dangerous.

_____

_____

2. Propose an explanation for why paramedics often use a mixture of oxygen and carbon dioxide instead of pure oxygen to resuscitate victims of carbon monoxide fumes.

_____

_____

3. Propose an explanation for why changes in altitude affect heart rate.

_____

_____

_____

4. Would you expect changes in heart rate to correlate with changes in respiration rate? If so, why?

_____

_____

_____

5. The longest periods underwater in near-drowning incidents occur in cold water. Propose an explanation for why recovery is possible after such extended periods without oxygen.

_____

_____

_____

6. Can you think of any examples of plant homeostasis? If so, describe them.

_____

_____

_____

7. Why is homeostasis important to organisms?

_____

_____

_____

8. Can you think of any homeostatic relationships in nonliving things? If so, describe them.

_____

_____

_____

9. About how many times does a human heart beat in an average lifetime? List all the assumptions you had to make to calculate your answer.

_____

_____

_____

_____

Investigation 7 *What Ways Can Isopods Sense Their Environment?*

Introduction

You may have noticed "little gray bugs" living under plants and rocks in wet places. Some people call these pill bugs; others call them wood lice or sow bugs. A biologist would call them *isopods*. Isopods are part of a large group of animals called *crustaceans*, which includes such better known members as crabs, shrimp, crayfish, and lobsters.

If you were to look closely at an isopod, you would see that it has many legs, a segmented body, and two pairs of antennae. Does an isopod have eyes, ears, a nose? Certainly, they move about and appear to know where they are going. Direct observation, however, does not reveal such sense organs as these. How, then, might isopods sense their environment? In this investigation, you will have the opportunity to design experiments and gather data to find out whether isopods can see, smell, or hear.

Objectives

1. To design controlled experiments that produce quantitative data that indicate ways in which isopods sense their environment.

2. To generate and analyze quantitative data that separate responses due to chance from responses due to a specific cause.

3. To state alternative hypotheses and communicate the reasoning used to systematically test them.

Materials

10–15 isopods (per group)	wire loops
bran or breakfast cereal	scissors
source of moisture (apple slice)	straws
hand lenses	rulers
light sources	powder
plastic wrap	shoe box (with a cover you make)
colored cellophane	chemicals with strong odors (e.g., vinegar, ammonia, perfume)
masking tape	ice
white paper	heat source (alcohol burner)
colored paper	wood splints

Procedure

1. Work with one or two lab partners. Obtain at least five isopods from the cultures and spend a few minutes observing and recording their behavior. Record your observations on a data sheet. **Caution: Use care when handling live animals.**

2. Spend 15 to 20 minutes observing and recording the animals' responses to various stimuli suggested by the materials list.

3. Design an experiment to answer a question that requires observation of the animal's reaction to some external stimulus (*e.g., Can isopods see? Can isopods smell? Can isopods hear? Can isopods communicate?*). **Caution: Be careful if you use a heat source or strong chemicals in your experiment. Wear safety goggles if using ammonia. Do not taste any chemicals. Note that ammonia is poisonous.**

You probably will need to obtain more isopods for your experiment. Try to obtain quantitative data instead of only the qualitative data you have gathered to this point. In other words, just because one isopod reacts in a particular way does not mean they all will. Think about how you will obtain general results.

4. Be prepared to discuss your design with your teacher or in a class discussion.

5. What is the independent variable tested in your experiment?

6. What variables should be held constant?

7. What is the dependent variable recorded?

8. Keep a record of your procedure, observations, and conclusions on your data sheet. Use the following format:
   I. Question Asked
   II. Alternative Hypotheses (assumed causes)*
   III. Procedure (experimental design)
   IV. Expected Result** (prediction)
   V. Actual Result (data, observations, tables)
   VI. Discussion and Conclusions
      * Suppose, for example, that in response to the question, *How do isopods locate food?*, you guess that isopods use their sense of sight to locate food. Your hypothesis is, *Isopods locate food by using their sense of sight.*
      ** If we assume this hypothesis is correct and put some isopods in a dark box with food at one end and some isopods in a lighted box with food at one end, then *we would expect that the isopods in the lighted box will find the food but that those in the dark box will not.* The italicized portion of the previous sentence is the expected result (what should happen in your experiment, assuming the hypothesis is correct).

9. Be prepared to communicate your findings to the class.

Study Questions

1. A biology student wanted to test the response of mealworms to light and moisture. To do this, she set up four boxes as shown. She used neon lamps for light sources and constantly watered pieces of paper in the boxes for moisture. In the center of each box, she placed 20 mealworms. One day later she returned to count the number of mealworms that had crawled to both ends of each box.

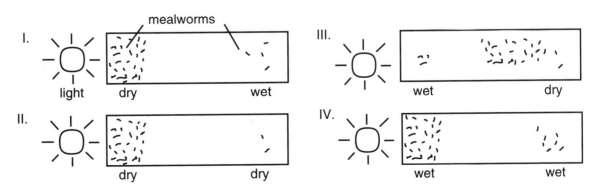

a. What two variables are being tested by the student in this experiment?

_____

_____

b. Name two values for each variable.

_____

_____

c. Name two variables that should be held constant to effectively test the effect of moisture and light.

_____

_____

d. Which of the four boxes (taken by themselves) represent controlled experiments?

_____

_____

e. Look only at Box I. What conclusion can be drawn from these data regarding mealworms' response to light? Explain. The worms' response to moisture? Explain.

_____

_____

_____

_____

f. Look only at Box II. What conclusion can be drawn from these data regarding mealworms' response to light? Explain. To moisture? Explain.

_____

_____

_____

_____

g. Knowing what you now know from observing Box II, what does Box III tell you about mealworms' response to moisture? Explain.

_____

_____

h. Suppose Box IV had only four mealworms in it and three of them were near the light end and one was near the dark end of the box. Also suppose that Box IV was the only box included in the experiment. Would you conclude from these data that mealworms in general prefer light over dark conditions? Explain.

_____

_____

2. Think of a question from your experience that would require conducting a controlled experiment to answer. For example, do the new solid-centered golf balls really travel farther than the old liquid-centered ones? State your question and describe your experimental design. What are the independent and dependent variables? What variables must be held constant? How can you be sure any observed differences are not due to chance?

_____

_____

_____

_____

_____

Additional Activity

The next series of questions deal with investigations of planarian behavior. The following materials will be needed:

planaria	marking pen
dechlorinated water	light source
test tubes, 13 x 100 mm	ice
test-tube rack	heat source
dissecting needle	raw liver
dropper	various foods
cork stopper	vinegar
ammonia	salt water, sugar water

1. Design controlled experiments to answer these questions: _Is a planarian's response to light is positive or negative? Is the response to be the same under all light intensities? What receptors are involved in the planarian's response to light?_ Test your hypotheses.

2. Design a controlled experiment in which you can observe the combined effects of gravity and light on the planarian.

3. Since light produces heat, can you be sure the planarian is responding to light and not to heat? How would you set up an experiment to discriminate between these two variables?

4. Set up an experiment in which you can observe and describe planarian feeding behavior. Can you provoke feeding behavior with a stimulus other than food? How would you determine what parts of the planarian's anatomy are used in sensing and reacting to the food?

5. You can test a planarian's response to chemicals. Your teacher will help you set up the experiment. Use more than one planarian to see whether they respond the same way. Try to explain your observations.

Reasoning Module 3 *Causal Relationships*

Introduction

I once knew a woman who was married six times. Her first husband was killed when his automobile was struck by a truck. Her second husband was struck down in his prime by a sudden and unexpected heart attack. Her third husband drowned while swimming in the ocean. Her fourth husband met his untimely demise when he choked on a piece of steak. Her fifth husband contracted a rare and exotic disease and died while on a trip to South America. Her sixth husband slipped on a roller skate that had been left at the head of a long flight of stairs. The slip resulted in a fatal tumble to the bottom.

Question: Would you marry this woman?

Sometimes a series of events just stretches the laws of probability too far. When this occurs we assume a hidden cause.

Objective

To further understand probability and its relationship to causality.

Materials

sack with cardboard mealworms

Procedure

1. Reading the first essay and answer the study questions.

2. Read the second essay and do the suggested activities.

3. Work through as many of the study questions as you feel necessary.

Essay 1: Two in a Row? You Must Be Kidding!

When we last left the two girls with the tennis balls, the second girl had guessed that the yellow ball was in the first girl's left hand. Was she correct? Actually, the yellow ball was in the right hand, so she was wrong.

At seeing this, the second girl exclaimed, "Oh, come on already. I'm in a hurry, so just give me the yellow ball."

To this, the first girl said, "No deal, but I will be nice and give you another try. But this time you must guess correctly twice in a row before I'll let you have the yellow ball."

What are the second girl's chances of guessing correctly twice in a row?

Obviously, the chances are less than 1 out of 2. It's tough to guess right once, but guessing right twice in a row is even tougher. Is it twice as hard? Let's see just how hard it is.

The sequence of events we want is "picking the yellow ball on the first try" and "picking the yellow ball on the second try." How many possible sequences of events are there? By using the following "tree" diagram, we can generate the possible sequences.

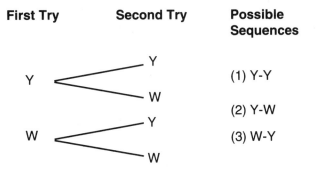

First Try	Second Try	Possible Sequences
Y	Y	(1) Y-Y
	W	(2) Y-W
W	Y	(3) W-Y
	W	

The diagram shows that on the first try the second girl could pick either the yellow (Y) or the white (W) ball. If she picked the yellow ball on the first try, she could pick either the yellow or the white ball on the second try. If she picked the white ball on the first try, she could also pick either the yellow or the white ball on the second try. So the sequence of possible events is Y and Y, Y and W, W and Y, W and W.

Four sequences are possible. Because the sequence of events we want (Y-Y) is 1 out of 4 possible sequences, the chances of obtaining this sequence is 1 out of 4, or 1/4. Again, if we extrapolate to 100 cases, we would predict she would be correct about 25 out of 100 (25%) of the time. We would say her chance of being correct twice in a row is about 25%.

Study Questions

1. What is the chance (probability) of tossing a coin twice and having it come up heads both times?

2. What is the probability of rolling a die twice and having it come up with a 6 both times?

3. Suppose four balls are put into a sack. One of the balls is red, and the other three are blue. What is the probability of reaching in and picking a blue ball on the first try? If you put the ball back into the sack after each pick, what are your chances of picking a blue ball again?

4. A newly married couple proclaimed that they were planning to have children right away. They both agreed that they wanted one girl and one boy. Assuming they will have two children, what are their chances of having one girl and one boy?

5. Local telephone numbers contain seven digits. Suppose you wanted to call a friend but you forgot the last two digits of his phone number. What is the probability of your correctly guessing the last two digits on the first try?

Essay 2: The Mealworms in the Box

When asked to investigate the behavior of mealworms, Carlo decided to find out whether mealworms respond to light. He put four worms into the center of a box that had a neon light shining at one end. The other end was covered with black paper to shade it. Carlo figured that if the mealworms "liked" the light, the four worms would move to the lighted end. If they "liked" the dark, they would crawl to the dark end. Fifteen minutes after putting the mealworms into the center of the box, he checked to see where the worms had crawled.

The diagram shows the results.

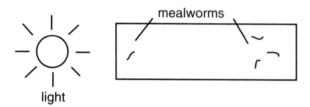

As you can see, one of the worms is in the light area, and three of the worms are in the dark area. Does this mean that the worms prefer the dark? If so, why is one worm still in the light? Maybe it is just a strange worm. Or maybe it's not strange. Maybe Carlo's results are just due to "chance." Maybe if he waits awhile, all the worms will crawl to the lighted end. In other words, what is the probability that this result is due to chance alone and not due to the light?

How does one interpret such results as these? This is the central question in analyzing scientific data. To understand the answer, we need to understand more about probability.

Do the following activity as a way of better understanding the situation.

Pick up a sack containing pictures of Carlo's four mealworms. Notice that each mealworm has two sides, one marked L for light and one marked D for dark. Put the four mealworms back into the sack and shake them up.

Turn the sack over and dump the worms onto the table. How many worms have L's showing? How many have D's showing?

Imagine that those with the L's showing are the ones that crawled to the lighted side of Carlo's box. The ones with D's showing are the ones that crawled to the dark side of the box. Record these numbers on your data table (see example) next to trial #1.

Put the four worms back into the sack and shake them up again. Dump the worms out onto the table again and record on the data table next to trial #2 the number of L's and D's that turned up this time.

Repeat this procedure 30 more times.

Trial #	Number of Worms in Light Side/Dark Side	Trial #	Number of Worms in Light Side/Dark Side
1		17	
2		18	
3		19	
4		20	
5		21	
6		22	
7		23	
8		24	
9		25	
10		26	
11		27	
12		28	
13		29	
14		30	
15		31	
16		32	

How many times did 4 L's and 0 D turn up?

_____

How many times did 3 L's and 1 D turn up?

_____

How many times did 2 L's and 2 D's turn up?

_____

How many times did 1 L and 3 D's turn up?

_____

How many times did 0 L's and 4 D's turn up?

_____

    If these numbers were plotted on a frequency graph, what shape would the curve be? Why do you think the curve has that shape?

_____

_____

The numbers you obtained represent the outcome of an actual experiment. Compare those numbers with what would be expected to occur according to a tree diagram. The tree diagram would look like this:

First Worm	Second Worm	Third Worm	Fourth Worm	Possible Sequences

				(1) L-L-L-L
				(2) L-L-L-D
				(3) L-L-D-L
				(4) L-L-D-D
				(5) L-D-L-L
				(6) L-D-L-D
				(7) L-D-D-L
				(8) L-D-D-D
				(9) D-L-L-L
				(10) D-L-L-D
				(11) D-L-D-L
				(12) D-L-D-D
				(13) D-D-L-L
				(14) D-D-L-D
				(15) D-D-D-L
				(16) D-D-D-D

The diagram indicates that the first worm could have gone to either the light side (L) or the dark side (D). Suppose it went to the light side. The second worm could then have gone to either the light side or the dark side. Suppose the second worm went to the light side. The third worm could then have gone to either the light side or the dark side, and so on.

As you can see, 16 sequences or ways are possible in which the worms could have arranged themselves. Of these 16, only 4 have one worm in the light and three in the dark (numbers 8, 12, 14, and 15). So the probability of Carlo's result occurring just by chance is 4 out of 16, or 1 out of 4. In percentages, this would be 25 out of 100, or 25%.

This is not a very large chance, so you could conclude that the dark (or absence of the light) was probably the cause of the worms being in that side of the box.

Notice, however, that you can't be certain about this. If you conclude that mealworms prefer the dark, you could be wrong. The worms actually may not respond to light or dark at all. If you did the experiment four times, you would expect this result in 1 out of the 4 experiments just by chance alone.

Study Questions

1. An experimenter wanted to find out whether rats can smell cheese from 30 cm away, so she set up a T maze as shown below. A piece of smelly cheese was placed behind a screen at one end of the T. The other end had a screen but no cheese. Both screens were 30 cm away from the middle of the T.

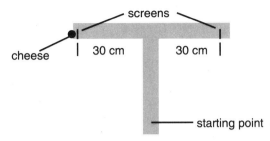

Five rats were used in the experiment. Each one started from the starting point and walked along the maze until it reached the junction. Three of the rats turned to the left and found the cheese, and two rats turned to the right.

What is the probability of this result occurring due to chance alone? From this experiment, would you conclude that rats can smell cheese from 30 cm away? Explain.

_____

_____

_____

2. Recently at the Valley Shopping Center, a man was giving a group of shoppers the "Cola Challenge." The Cola Challenge amounts to giving people a sip of Brand X and a sip of Brand Y without them knowing which is which. The people then report which tastes better to them.

   While I watched, I noticed that three people said Brand X tasted better. One person said she liked Brand Y better, and two people said they couldn't tell the difference.

   a. What fraction of the people preferred Brand X?

   _____

   b. Suppose the Cola Challenge is given to one more person. What would you say the chances are that the person will prefer Brand X?

   _____

   _____

   _____

   c. Does this sample of data show that people in general prefer Brand X to Brand Y, or is the result just due to chance? Explain.

   _____

   _____

   _____

3. At a dinner party for 20 people, apple, cherry, and banana cream pies were served for dessert. On the day following the party, 4 of the 5 people who ate banana cream pie came down with severe stomach pains and nausea.

    a. Does it seem likely that the banana cream pie was the cause of the stomach pains and nausea? Explain.

_____

_____

_____

    b. Can you figure out the probability of this many people getting sick due to chance alone? If not, what other information do you need?

_____

_____

_____

Brain Bender 1

How many different license plates can there be for your state? Generate a formula for finding the number of possible combinations.

Brain Bender 2

You are on a television game show and are given a choice of three doors. Behind one door is a new car; behind the others are pigs. Suppose you pick Door 1, and the host, who knows what is behind each door, opens Door 2 or 3, whichever one has a pig behind it. He then asks whether you want to switch your pick to the remaining closed door. Should you switch? Why? Why not? (Hint: You may be thinking that because two doors remain unopened with a car behind one and a pig behind the other, your chances are 1 out of 2 for each door. Although to most people this seems reasonable, it is wrong. Generate all possible combinations of results before you jump to a hasty and incorrect conclusion.)

Investigation 8 *What Is Inside a Frog?*

Introduction

Vertebrates, like all life forms, are faced with two basic challenges: obtaining and using energy, and reproducing. In this investigation, you will examine the internal anatomy of the frog to discover structures involved in these functions. Imagine that you are the first person ever to dissect a frog. Your main job will be to discover what structures exist, to observe their characteristics and their connections with other structure, and to try to figure out what each structure does.

Objectives

1. To observe the internal anatomy of a representative vertebrate.
2. To discover the internal structures responsible for energy acquisition and use, as well as those involved in reproduction.
3. To create alternative hypotheses regarding organ functions.
4. To propose experiments to test the alternatives.

Materials

dissection kit
dissecting pan
preserved frog
dissecting microscope

Procedure

1. Through careful dissection, observation, and creative hypothesis formation, attempt to answer these questions: *What structures are found inside a frog? Which structures are connected to one another, and in what way(s) are they connected? What does each structure do?*

2. Working with a partner, obtain a dissecting kit, a pan, and a frog. **Caution: Be sure to wear your safety goggles and lab coat during this investigation.**

3. Open the ventral side of the frog. **Caution: Be careful when using sharp dissecting tools. Also be aware that internal structures lie very close to the surface.**

4. Once you have cut and pinned back the ventral layer of skin, note such characteristics as the size, shape, color, and points of attachment of the organs you find. Keep in mind that your job is to locate specific organs and to speculate about their possible function(s). You may or may not be able to name the organs. At this point, names are not important. Use the organs' characteristics and their connections with other organs as the basis for generating hypothesized functions. On a separate sheet of paper, sketch your observations and label the organs with names or letters.

5. Record in a table the hypothesized functions and reasons you believe the organs function in those ways. Be prepared to discuss your ideas in a class discussion.

6. Using extreme care, remove organs observed thus far to reveal more dorsally located organs. Sketch these and add them to your table, along with hypothesized functions and reasons.

7. Locate and carefully remove all the structures you think the frog uses to process food.

8. Place removed structures off to the side of the pan and measure and record the total length of these structures.

9. Draw a diagram of the food-processing structures you removed. Show the sequence of the structures and the points of attachment for each. What do you think is the function, based on your observations, of each structure? Add this information to your table if you have not already done so.

10. Use a dissecting microscope to take a closer look at the internal and external lining of these food-processing structures. Do you see differences? If so, what are they, and why do you think they exist? Find out what this frog ate.

_____

_____

11. You may wish also to cut through the dorsal surface of the frog to locate its brain. Ask your teacher for advice as this is a very delicate procedure. Feel free to make any additional cuts that may be of interest to you, keeping in mind that the frog was once a living, breathing organism.

Study Questions

1. Trace the passage of a fly through the digestive system of a frog. Include the names and functions of all organs involved.

_____

_____

2. Cite and explain two ways in which surface area is increased in the digestive system of a frog.

_____

_____

3. Distinguish form and function of arteries, veins, and capillaries.

_____

_____

4. Describe the interrelationship between the circulatory and respiratory systems.

_____

_____

5. Describe the interrelationship among the circulatory, digestive, and urogenital systems.

_____

_____

6. Define organ, system, and organism. How are they similar? Different? How are they related to one another?

_____

_____

7. Select one organ listed in your table. Describe an experiment that could be performed on living frogs to test its hypothesized function.

_____

_____

Investigation 9 *What Do Living and Nonliving Things Looks Like Under the Microscope?*

Introduction

Advances in all branches of science have been brought about by the development of tools and instruments capable of extending human senses. The compound microscope is one such tool. In this exploration, you will examine a number of living and nonliving objects under the microscope in an attempt to discover microscopic differences between living and nonliving things. These observations should reveal fundamental differences between living and nonliving things.

Objectives

1. To recognize structural patterns of living and nonliving things at the microscopic level.

2. To gain experience in using a microscope to gather scientific information.

Materials

compound microscope   lens paper
slides and coverslips   scalpel or razor blade
methyl cellulose   dropper

Living Objects	*Once-Living Objects*	*Never-Living Objects*
algae culture	tomato	water
pond water	potato	mud and water
elodea	bamboo shavings	salt
onion root tip slide	cork	sand
coleus	snake skin	synthetic sponge

Procedure

1. With a partner, obtain a microscope and prepare it for viewing.

2. Obtain at least three samples from each category of objects (living, once-living, never-living) and examine them under the compound microscope to answer the question, *How are living, once-living, and never-living objects similar to and different from each other?* **Caution: Never taste or eat any plants or plant parts.**

_____

_____

3. When making slides, use wet mounts only when appropriate; some materials may need only dry mounts. **Caution: Handle glassware carefully to avoid breakage.**

4. Record observations on your data sheet. Be sure to include the name of each specimen, the magnification used, and its category (living, once-living, or never-living).

5. Be prepared to discuss your observations, classifications, and arguments with the class.

Study Questions

1. What fundamental microscopic differences exist between living and never-living objects?

2. How does the microscopic image compare with the actual orientation of the object on the slide?

3. What sort of evidence would be needed to support the generalization that all organisms, including humans, are composed of cells? Do you know of any organisms that are not composed of cells? If so, which one(s)?

4. Not all cells are the same size and shape. What might be some reasons for these differences?

5. Did your observations reveal any consistent differences between plant and animal cells? If so, what are they?

Investigation 10 *How Does Cell Structure Relate to Function?*

Introduction

You might have noticed that buildings are designed differently, depending on their use. A house is designed differently from, say, a theater. Also, buildings are designed differently depending on where they are located. A house in Alaska, for example, requires more insulation than one in Hawaii.

Are similar differences found in the cells of living things? Cells in multicellular organisms are located in many different places. Are they also designed differently? If so, in what ways? How do these differences relate to different functions? In this investigation, you will observe a number of different types of cells, note the structural characteristics, and try to infer possible functions. In other words, you are going to explore the relationship between cell structure and function.

Objectives

1. To observe the structure of cells and attempt to create alternative hypotheses about their functions.

2. To note the wide diversity among cell types.

Materials

compound microscope
unlabeled slides of a variety of animal cells
unlabeled slides of a variety of plant cells
carrot slice
dish of water

Procedure

1.   With a partner, observe under the microscope the ten numbered specimens. **Caution: Handle the slides carefully to avoid breakage**.

2. For each cell type, try to answer the question, *What is the function of this type of cell?* Note the cell size, shape, and any other characteristics that might provide clues about cell function. Speculate as to cell function based on structure and relationships with other cells on the slide. Make drawings of your observations on a data sheet that includes slide number, appearance, possible function(s), and your reasons for the hypothesized functions.

3. Join another pair of students to compare your ideas. Record reasons for agreement or disagreement and be prepared to discuss your ideas and arguments in a class discussion.

4. During a class discussion, your teacher will tell you where each specimen came from. Is this information consistent or inconsistent with your hypothesized functions? Explain for each specimen.

5. If time permits, obtain a slice of carrot (cross-sectional) and place it in a dish of water. Use a hand lens to observe the slice. Try to identify as many types of tissues as you can. Draw them on a separate sheet of paper. Create alternative hypotheses about possible functions. Store the cross section overnight in a cool location. On the next day, observe your carrot cross section again. Do your observations support or contradict your original hypotheses? Explain.

Study Questions

1. How do cells differ from tissues?

   _____

   _____

2. List ways in which cells differ.

   _____

   _____

   _____

   _____

3. List functions that might be accomplished best by cells with the characteristics listed in question 2. Provide a reason for each hypothesized structure-function relationship.

   _____

   _____

   _____

   _____

   _____

   _____

   _____

Investigation 11 *What Is Inside Cells?*

Introduction

You have discovered that organisms are composed of one or more cells. Like you, early researchers saw cells as bodies filled with shadowy, gray material. They gave this mysterious material the name *protoplasm,* but they really had little idea what it was like and how it functioned. It was not until the invention of the electron microscope in 1933 that biologists were able to magnify cells enough to see that some cells were filled with tiny substructures. Biologists were able then to describe the appearance of these substructures--or *organelles,* as they were named--and to begin to create and experimentally test alternative hypotheses about their functions.

In this investigation, you will view photographs, taken through an electron microscope, of various types of cells to explore organelle structure. Using these photographs (called electron micrographs), along with information derived from past experiments, you will explore the structure and function of a variety of cell organelles.

Objectives

1. To explore, compare, and contrast the nature of the protoplasm in plant, animal, blue-green algal, and bacterial cells.

2. To observe and describe the structure of cell organelles and to create and discuss alternative hypotheses about their functions.

Materials

electron micrographs
markers or crayons
butcher paper
historical information about cell organelles

Procedure

1. Working with a partner, observe the set of five electron micrographs. They include a euglenoid, a plant cell (spinach), an animal cell (lung), a bacterial cell, and a blue-green algal cell. *What similarities and differences do you observe among them? Are there any inner-cell structures common to more than one of the cells?* If so, describe these common structures.

_____

_____

2. Now look at either the plant cell, the euglenoid, or the animal cell. Draw a large picture of it on butcher paper. Include at least one of each of the different organelles that you can see.

3. In a class discussion, compare your drawing with those of other groups. Your teacher will provide names for the organelles that were found.

4. Take a closer look at each organelle by visiting the stations that have been set up around the room. At the stations are several micrographs of a particular organelle and a list of discoveries made by cell biologists about its function. While keeping in mind the basic functions of all organisms, propose alternative hypotheses about the function of each cell organelle. In other words, the question you are trying to answer is, *What is the function(s) of each cell organelle?* Record your hypotheses.

5. Compare your hypotheses with the historical information about the organelles. Does this information conflict with predictions derived from any of your hypotheses? If so, revise your hypotheses to fit the information.

6. After you have visited each of the stations, compare your hypotheses and observations with those of your classmates in a class discussion. Revise your list of hypothesized functions as it becomes necessary.

Study Questions

1. Why is it necessary to use electron micrographs to observe cell organelles?

2. Many processes that occur in cells are associated with membranes. What evidence, if any, can you offer that supports this statement?

3. Biologists think that a relationship exists among the nuclear membrane, the Golgi body, the endoplasmic reticulum (ER), and the cell membranes. What evidence, if any, have you obtained to support this idea?

4. Explain how ER structure supports the idea that form is related to function.

5. Some species of algae have a single mitochondrion in each cell, while muscle cells may each contain as many as a thousand mitochondria. Suggest a possible explanation for this difference.

6. Mitochondria and lysosomes can move freely inside the cell. Why might this be an advantage?

7. Many of the chemical reactions that occur in mitochondria occur on a membrane. What might be an advantage of the infolding of the inner membrane?

8. The islets of Langerhans are a group of cells in the pancreas that produce the hormone called insulin. Which organelle would you expect to be very abundant in these cells? Why?

_____

_____

9. What similarities exist between the functions of the cell organelles and the functions of organs within an organism?

_____

_____

10. Some organelles are needed only temporarily. They are then broken down into molecules that can be recycled. Which organelle(s) might be involved in this process? Explain your choice(s).

_____

_____

Investigation 12 *What Gas Do Growing Yeast Cells Produce?*

Introduction

Yeast cells grow and reproduce in bread dough. As they grow and reproduce, they give off a gas that is trapped and forms bubbles in the dough, causing the dough to expand. Do you know what the gas is? Is it oxygen? Is it carbon dioxide? Or is it something else? And precisely what are the right conditions for yeast growth? Do yeast need a food source as animals do, or are they plants that use light energy to produce their own food? In this investigation, you will attempt to grow yeast cells and will experiment to try to find answers to these questions.

Objectives

1. To microscopically observe yeast cell reproduction.

2. To determine through experimentation the conditions necessary for yeast reproduction and population growth.

3. To test alternative hypotheses to identify the yeasts' food/energy source and the gas given off.

Materials

flexible straws
screw-top test tubes
bromthymol blue (BTB) solution
compound microscope
depression microscope slides and coverslips
dry baker's yeast
table sugar (sucrose) 5% solution
1-hole rubber stoppers
250-ml glass beakers
S-shaped glass tubes
grow lights

Procedure

1. Can yeast cells produce their own food, using the energy of light, or do they require a food source such as sugar? What ideas do you have at this time? What reasons can you give for these ideas? Record your ideas and reasons.

_____

_____

_____

2. Use the materials provided to design and conduct an experiment to answer the questions in step 1. Record your design and results.

_____

_____

3. What gases might yeast cells take in and expel? Record some possibilities.

_____

    a. If you assume that the hypothesis that yeast cells take in O_2 and expel CO_2 is correct, what color should growing yeast cells cause a blue BTB solution to become? A yellow BTB solution?

    _____

    _____

    b. If you assume that the hypothesis that yeast cells take in CO_2 and expel O_2 is correct, what color should a blue BTB solution become? A yellow BTB solution?

    _____

    _____

4. Use the materials provided to design and conduct an experiment to test the above alternative hypotheses. Record your design and results.

_____

_____

_____

5. Use a compound microscope to observe a wet-mount slide of yeast culture.

_____

_____

6. Sketch observed details of the yeast cells as seen through the microscope. Do any of the cells appear to be reproducing? Explain.

_____

_____

_____

_____

_____

_____

7. Identify with labels the parts of the yeast you can recognize. Be prepared to report your observations in a class discussion.

Gas-Collection Apparatus

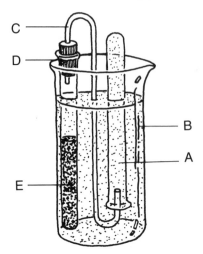

To collect the gas produced by yeast:

a. Fill the collection test tube (A) with BTB indicator solution and invert the tube in the 250-mL beaker.

b. Fill the 250-mL beaker (B) with 200 mL of water and BTB indicator.

c. Insert one end of the S-shaped glass collection tubing glass (C) into a 1-hole rubber stopper (D) and the other end into the water filled collection test tube (A). **Caution: Glass tubing is fragile. You may needs your teacher's assistance on this step.**

d. Fill the culture tube half full of yeast culture (E).

e. Place the 1-hole rubber stopper and glass tubing on the culture test tube to seal the system.

Study Questions

1. What evidence exists to support the hypothesis that individual yeast cells take in O_2 and expel CO_2? That they take in CO_2 and expel O_2?

2. What were the independent and dependent variables in your experiment? Which variables should have been held constant? Why?

3. Did you conduct a controlled experiment? Explain.

4. In light of your results, would you consider yeast a plant? An animal? Neither? Both? Explain.

Reasoning Module 4 *Controlling Variables*

Introduction

One day at the lake, two boys were overheard arguing about who could do more push-ups. The bigger boy said, "There is no way you can do more push-ups than me. I'm bigger and stronger than you."

To this, the smaller boy replied, "OK, let's see you prove it. You go first."

The bigger boy immediately got down on the sand and did 25 push-ups before he tired and could do no more. The smaller boy watched patiently and appeared not the least bit shaken by the bigger boy's performance.

At the conclusion of the bigger boy's effort, the smaller boy smiled and jumped into the water up to his calves. He got down in the water and proceeded to do 26 push-ups without breathing deeply. When he finished, the bigger boy protested, "Hey, you can't do that. That's not fair. It's a lot easier to do push-ups in the water than on the sand!"

What do you think? Was it fair? Obviously not. One boy doing push-ups on the sand and the other in the water is not a fair test to find out who can do more push-ups. What would make it fair?

In this activity, you will explore how to conduct fair tests and why fair tests are so important in science.

Objective

To recognize, design, and conduct controlled experiments (fair tests).

Materials

Whirlybird® base consisting of 2 pieces
arm
post with bolt and wing nut
6 rivets
3 rubber bands
screw eye
timer

Procedure

1. Begin by reading the first essay.

2. Do the experiments suggested in the second essay, and complete the study questions.

Essay 1: The Tennis Balls Again

As the girl with the white tennis ball continued down the bumpy dirt road, she noticed a boy walking toward her. The boy was bouncing an orange tennis ball. At seeing the bouncing orange ball, she noted that it seemed to be bouncing a lot higher than her white ball.

After saying hello, the boy replied, "Hey, your tennis ball is a real dud. My orange ball is a lot bouncier than yours."

Not wanting to be outdone, the girl responded, "Oh, no, it's not. Mine is bouncier than yours and I can prove it."

To this, the boy said, "Oh yeah? Let's see you prove it. Go ahead. Drop both balls and let's see which bounces higher."

At this, the girl held her white ball over her head and the boy's orange ball near her waist and dropped them both at the same time. The white ball bounced higher.

To this, the boy protested, "Hey, that's not fair. You can't drop them from different heights." So the girl took the two balls again and dropped them from the same height. But this time she dropped them so that the white ball hit a hard spot in the road and the orange ball hit a soft spot. Again the white ball bounced higher. Again the boy protested, "That's still not a fair test. My ball hit a soft spot and yours hit a hard spot. Do it again but this time don't drop them from different heights and don't let one hit a soft spot."

So again the girl held up the two balls and released them. She did what the boy told her, but this time she released them so that the white ball hit the sidewalk and the orange ball hit the road. Again the white ball bounced higher than the orange ball.

By this time the boy was getting rather upset. Again he protested, "Don't drop them from different heights. Don't let one hit a hard spot and the other a soft spot, and don't let one hit the sidewalk."

"OK, OK," exclaimed the girl. "Let me try again." So again she held up the two balls and released them from the same height. They both hit hard spots in the road. But again the white ball bounced higher. This time she had cleverly spun the orange ball as she dropped it so that when it hit the road, it bounced at an odd angle and did not rise very high into the air.

At seeing this, the boy was so upset at the girl's failure to conduct a fair test that he grabbed his orange ball, turned around, and went off down the road, muttering to himself.

Suppose you were the boy. What could you have said to the girl to keep her from messing up the test? You could, of course, do what the boy did—tell the girl a number of things that she should *not* do (do not release the balls from different heights, do not let one hit a soft spot while the other hits a hard spot, do not let one hit the sidewalk while the other hits the road, do not spin one ball and not the other). If you did this, however, your sentence would get very long and she still might fool you. A better way is simply to tell the girl what she *should* do. And that is, *do the same thing to both balls.*

If she did the same thing to both balls (released them from the same height, let them both hit the same road surface, dropped them both with no spin), she would have conducted a fair test. Tests have to be fair or else you cannot be sure whether the results are due to the variable under consideration (in this case the difference in the two balls) or due to some misleading variable (the height of release, the surface the balls hit, the spin). Fair tests are so important in science that they have been given a special name. They are called *controlled experiments.* "Controlled" refers to the fact that all the variables that might make a difference (except the variable under consideration) are kept the same.

If you do a controlled experiment in which only one variable is different (in this instance, the balls themselves) and you get a difference in the result (in this instance, one ball bounces higher than the other), then you can be sure that the difference is due to that variable (the orange ball really is bouncier). It must be due to that variable because all other variables were the same (both dropped from same height, hit the same surface, etc.).

Essay 2: The Whirlybird ®

Let's see whether you can conduct some fair tests (controlled experiments) to identify variables that make a difference in the number of times a Whirlybird spins around before stopping.

Obtain the materials, and spend a few minutes attempting to assemble them so that the arm will spin around.

The diagram below shows one way to assemble the Whirlybird. By attaching one end of a rubber band to the screw eye and looping the other end over the staple in the post, the rubber band can be wound around the post by rotating the arm. On releasing the arm, the rubber band will fly off the staple and the arm will continue to spin around several times.

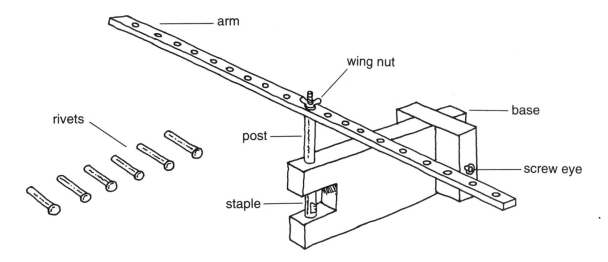

What variables do you think may affect the length of time the arm will spin? List the variables.

Carry out a series of controlled experiments to find out whether the variables you have listed really do make a difference. Record important information about your experiments and their results. To do so, you may wish to use copies of the following sample record sheet.

Sample Record Sheet
    Variable tested _____

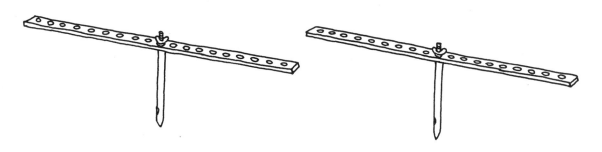

A. Number of rivets _____
    Number of rubber bands _____
    _____ twists of the rubber band
    Additional variables:

    _____
    _____
    Show where you placed the rivets.

B. Number of rivets _____
    Number of rubber bands _____
    _____ twists of the rubber band.
    Additional variables:

    _____
    _____
    Show where you placed the rivets.

    Does your experiment show that the variable tested really does affect the number of turns of the arm? _____
    Explain. _____
    _____
    _____

In your experiments, the number of times the arm spins is called the *dependent variable* because it is dependent on a number of other variables (number of twists of the rubber band, position of the rivets, etc.). These other variables are really the *causes,* while the change in the dependent variable is the *effect.* These causes are independent, or separate, of each other; hence, they are called *independent variables.* A controlled experiment is one in which a correlation is sought between an independent variable and the dependent variable when all the other independent variables are held constant.

Study Questions

1. For each of your experiments, can you be sure that any observed difference in the results was due to the variable being tested? Explain.

2. What was the dependent variable in the tennis ball experiments of the first essay?

3. What were the independent variables in the tennis ball experiments?

4. What was the dependent variable in the story about the push-ups?

5. What important independent variable was not controlled (kept the same) by the smaller boy in the push-up story?

6. Fifty pieces of various parts of plants were placed in each of five sealed jars of equal size under different conditions of color of light and temperature. At the start of the experiment, each jar contained 250 units of carbon dioxide. The amount of carbon dioxide in each jar at the end of the experiment is shown in the table.

   Which two jars would you select to find out whether temperature makes a difference in the amount of carbon dioxide used? Explain.

Experimental Conditions and Results

Jar	Plant Type	Plant Part	Color of Light	Temp. (°C)	Units CO_2 at End
1	willow	leaf	blue	10	200
2	maple	leaf	purple	23	50
3	willow	root	red	18	300
4	maple	stem	red	23	400
5	willow	leaf	blue	23	150

7. In Essay 3 of Reasoning Module 2, the swimming coach did an experiment to find out who were the fastest swimmers among the girls who tried out for the team.

   a. What was the experiment?

   _____

   _____

   b. Was it a controlled experiment? Explain.

   _____

   _____

   c. What was the dependent variable (effect)?

   _____

   _____

   d. What independent variables (causes) do you think are important in this situation?

   _____

   _____

8. Question 1 of the study problems in Reasoning Module 2 shows a graph of speed versus miles per gallon.

   a. What was the dependent variable (effect) in that situation?

   _____

   _____

   b. What was the primary independent variable (cause) investigated?

   _____

   _____

   c. What other independent variables might be important?

   _____

   _____

d. What important independent variable was controlled?

_____

_____

e. What important independent variable was not controlled?

_____

_____

f. How would you improve the experiment to find out whether a correlation really does exist between speed and miles per gallon?

_____

_____

Brain Bender

A student recently conducted an experiment to find out whether the color of the cup a person drinks from affects his or her perception of what he or she is drinking. The student predicted that people would prefer drinks from orange cups because his psychology teacher had remarked that orange was a warm color that had a soothing effect on one's disposition.

He tested this prediction by putting identical kinds and amounts of cola into two cups — one orange and one white. Wanting to do a controlled experiment, he proceeded in exactly the same way for each of the 15 persons he tested. First, he told them that they would be tested to find out which of two unknown kinds of cola they preferred. Each person was given 25 mL of the first unknown cola in a white cup. Then he or she was given crackers to clear out the taste. Next, each person was given 25 mL of the second unknown cola (actually the same kind of cola) in an orange cup. The student found that 11 out of the 15 people said that they preferred the cola in the white cup.

Questions:

a. Was this a controlled experiment? Explain.

_____

_____

b. Why do you think so many people chose the cola in the white cup?

_____

_____

c. Does this experiment prove that the color of the cup affects a person's perception of what he or she is drinking? Explain.

_____

_____

d. Would you suggest any changes to improve the experiment? If so, what are they?

_____

_____

Investigation 13 *How Do Multicellular Organisms Grow?*

Introduction

Two characteristics of living things are 1) they are composed of cells and 2) they grow. But how do multicellular organisms grow? Do their cells become larger? Does the amount of space between adjacent cells increase? Perhaps the organism increases its number of cells. Perhaps cells produce extracellular structures that serve as a sort of miniature skeletal system. Or perhaps growth occurs through some combination of the above.

In this investigation, you will have an opportunity to use the microscope to observe the cells of a rapidly growing portion of a plant (the root tip of an onion) to discover which, if any, of these alternative hypotheses are correct.

Objectives

1. To observe and record differences among cells in a growing organism.

2. To test alternative hypotheses that attempt to explain growth in multicellular organisms.

Materials

prepared slides of onion root tip and/or whitefish or ascaris cells	yarn
	red and white pipe cleaners
compound microscope	35-mm slides of meiotic sequence and slide projector
large sheet of paper	germinating seeds
scissors	waterproof marker
string	glue

Procedure

1. Observe the onion root tip slides under the microscope and try to answer the following questions: *What differences in cell size, structure, and spacing exist? What areas appear to be the youngest? The oldest? Do your observations support or refute any of your alternative hypotheses? If so, why?* **Caution: Handle glass slides carefully.**

_____

_____

_____

2. While observing, make clear drawings of any differences you see among the cells. Pay particular attention to possible differences in cell size in different locations of the root tip. Also note possible differences in threadlike structures in the cell nuclei. These are called *chromosomes*. Make as many drawings as you like. Label all obvious cell structures.

3. Copy each drawing onto a large sheet of paper. Make the drawings large and clear enough that all students can see them when the papers are hung around in the room.

4. Carefully observe all of the drawings.

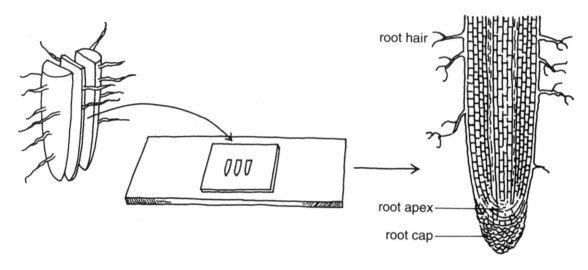

root hair

root apex

root cap

5. On the root tip slide, do you think you were looking at many types of cells or at one type of cell that was going through changes? Explain.

_____

_____

6. If you think the cells were going through changes, number your drawings in an order that might reflect the order of those changes. Sketch that sequence.

7. Which, if any, of the initial alternative hypotheses are supported by the class's observations? Explain.

_____

_____

_____

_____

8. What additional observations would be helpful to test your ideas?

_____

_____

9. Compare your hypothesized sequence of changes with the apparent sequence shown in the film. Does this comparison suggest that you need to modify your sequence? If so, what modifications are needed?

_____

_____

Study Questions

1. How long does mitosis take? The lifespan of a cell is the period from when it was formed to the time it completes division to form two new daughter cells. This period is called the cell's generation time. By observing living cells, biologists have determined the *generation time* for a variety of cell types. If we know the generation time, we can determine how long a cell spends in mitosis. Given that 100 cells are observed in the field under a microscope and 50 of these cells are undergoing mitosis and that the generation time is 2 hours, approximately how long does it take the cells to undergo mitosis? After you have arrived at an answer, reflect on your procedure. Try to write an equation that expresses

the numerical relationships using

N_o for total number of cells observed;
T_m for time spent in mitosis;
N_m for number of cells observed in mitosis; and
T_g for generation time.

2. Now that you have a method for estimating time spent in mitosis, try to apply that method to the data below for onion root tip cells.

N_m = 16 cells
N_o = 642 cells
T_g = 22 hours

   a. How long does an onion root tip cell spend in mitosis (T_m)? Show your calculations. Express your answer in hours and minutes.

   b. What fraction of an onion root tip cell's lifespan is spent in mitosis? Show your calculations.

   c. What percentage of an onion root tip cell's life is spent in mitosis? Show your calculations.

3. Observe replicating cells of a whitefish blastula or ascaris. Do any differences exist in the processes of cell replication between these animal cells and the onion cells? If so, briefly describe them.

4. Set up a cell model. Use string for cell and nuclear membranes; yarn for uncoiled, unreplicated chromosomes; and red and white pipe cleaners for replicated chromosomes. Move the cell parts around on your desk to simulate the process of mitosis. Practice so that you will be able to demonstrate the process to others.

5. You probably know you have the same number of chromosomes as each of your parents. Propose ways in which sperm and egg cells can be created so that their fusion results in a new individual whose cells have the same number of chromosomes as the parents. The actual process in which this occurs is called *meiosis*. View a set of 35-mm slides of meiosis or reduction division to answer the following questions:

   a. What happens to the number of chromosomes in the nucleus of the original cell?

   b. How many chromosome divisions occurred during sex-cell formation?

   c. How many new cells are formed at the end of this process?

   d. When during the process did chromosome replication take place?

   e. In the cells of living organisms, chromosomes occur in look-alike pairs called *homologous chromosomes*. Propose a possible reason for this.

6. Obtain some germinating seeds. Use a waterproof marker to place equally spaced marks along the length of the growing root. According to your hypotheses, most of the growth near the root tip occurs due to cell division, while most of the growth further back from the tip occurs due to cell elongation. If you assume that these hypotheses are correct, what would you predict will happen to the spaces between adjacent marks as the root continues to grow? Observe root growth for the next week or so to find out whether your predictions are confirmed.

7. The diagram on page 68 closely resembles an actual photograph of a set of replicated human chromosomes. Obtain a copy of the diagram from your teacher. Cut out the chromosomes, and try to arrange them in patterns. When you have arranged them in a pattern that makes sense to you, tape or glue them onto paper. Explain why you arranged this chromosome "karyotype" the way that you did.

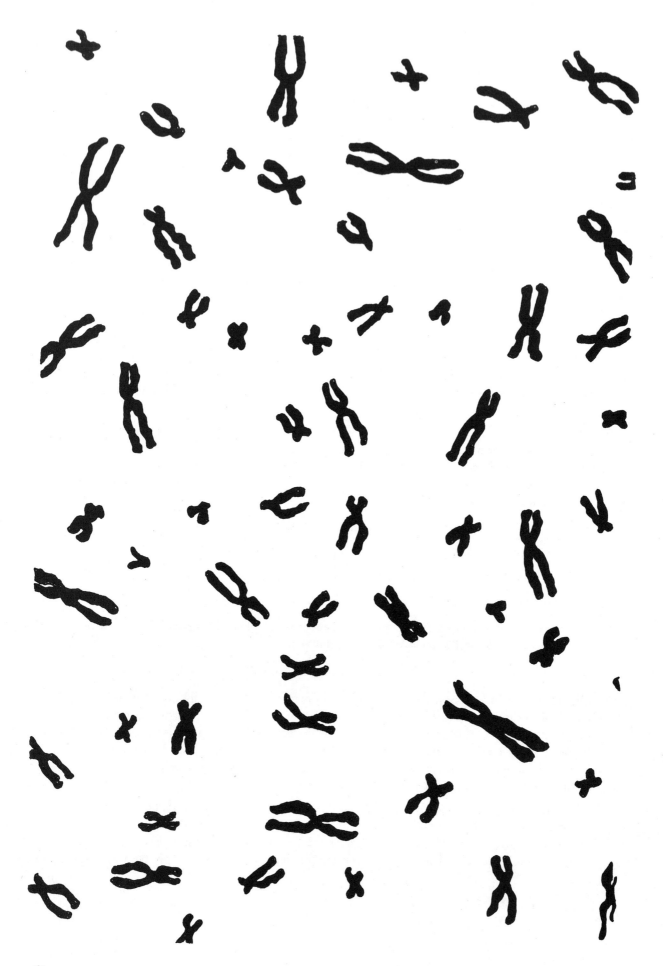

Reasoning Module 5 *Doing Science*

Introduction

Science is the attempt of people to understand the objects and events they experience in nature. People develop understanding about things they experience by asking questions and finding answers. What is life? What causes animals to die? What happens to frogs in the winter when their pond is frozen over? Why do so many different kinds of plants and animals exist? In attempting to find answers to such questions as these, you are doing science. Because all of these questions involve living things, finding answers to such questions involves doing biology — the science of life. Consider, for example, the following story about a horse named Clever Hans and the question, *How smart are nonhuman members of the animal kingdom?*

Clever Hans by all accounts was a very clever horse. According to newspapers around the turn of the century, he could identify musical intervals, understand Latin, and was quite good at math. When his owner would ask him to add numbers, Hans would tap out the answer with his hoof. For other questions, Hans would gesture with his head toward the appropriate pictures or objects.

Understandably, many people were skeptical about this, so a group of "experts," including two zoologists, a psychologist, a horse trainer, and a circus manager, were brought in to investigate. They watched closely as Hans's owner asked question after question to which Hans replied with near-perfect accuracy. Hans was even able to reply correctly to questions posed by perfect strangers. If the question called for the square root of 16, Hans would confidently tap four times with his hoof. The experts were unable to discover any tricks and thus were forced to conclude that Hans was a very clever horse indeed.

Do you agree with the experts? If not, how could Hans have correctly answered the questions? After reading the experts' report, a young psychologist proposed a different explanation. Suppose Hans was not able to think out the answers at all. Suppose instead he was able to monitor subtle changes in the questioner's facial expression, posture, and breathing that would occur when Hans arrived at the correct answer. Perhaps these cues could tell Hans when to stop tapping or moving his head.

How could this explanation be tested? The psychologist decided he needed to have some questioners who did not know the correct answers. If the questioner did not know the answers, then his or her expressions could not cue Hans, and Hans success rate should drop considerably. Sure enough, when the interrogator knew the answer, Hans succeeded on 9 out of 10 problems. When the interrogator did not know the answer, Hans's score dropped to just 1 out of 10. Hans had not learned math, music, or Latin after all. Instead, he had learned how to read people's faces and their body language. In one sense, Hans was not smart at all, but in another sense he was extremely perceptive.

Objective

1. To identify a pattern of scientific reasoning and investigation.

2. To recognize the differences among types of sentences based on the role they play in the process of doing science.

Materials

32 sentences written on two sheets of paper
scissors

Procedure

1. Obtain the sheets of paper containing the 32 sentences. The sentences are about 1) the candle-burning experiment from Investigation 1, 2) Clever Hans, the horse you just read about, and 3) a young child who has been waking up much too early in the morning. Cut the sheets of paper so that each sentence appears on a separate strip of paper.

2. As a homework assignment or in class with a partner, place the sentences face up on a table in front of you. Read each sentence and attempt to classify and arrange them in five to ten separate groups according to sentence type. You may want to start by grouping the sentences into three groups based on context (sentences about the candle-burning experiment, sentences about the horse, sentences about the child) and by arranging them in order to tell a story. Your final classification system, however, cannot be based on context. Rather, sentences of the same "type" regardless of context must by grouped together.

3. Be prepared to discuss your classification system with the rest of the class. You should be able to tell others the criteria you used: How are the various sentences similar? Different? What key words or phrases did you use, if any, as clues to sentence type?

4. Did all groups come up with the same system of classification? How many basic types of sentences are there? How can the sentences be combined to generate arguments that enable scientists to draw reasonable conclusions?

_____

_____

_____

_____

_____

_____

5. Read the essay "The Early Riser" and attempt to answer the Study Questions

Essay: "The Early Riser"

How does one do biology? How does one answer questions about life? How does one answer any question? Consider the following situation.

A few years ago, the behavior of a 1-year-old boy raised a question. The child was waking up at about five o'clock each morning. As far as his parents were concerned, this was too early. Why was he waking up so early? The problem was to discover the cause so that something could be done to get him to sleep longer.

Because the boy's awakening occurred in the summer when the sun was streaming through the window very early, his parents thought that perhaps the light was awakening him — a hypothesis. A _hypothesis_ is a tentative explanation put forth to account for some experience. It is a possible answer to the question raised. In this case, it is a tentative answer to the question, _What caused the child to wake up so early?_ A second hypothesis was that the child was hungry and that his hunger awakened him. Although other alternative hypotheses could be suggested, these seemed the most likely to the parents. Creating alternative hypotheses is an important first step in answering a question. The next step involves testing the hypotheses to find out which of the alternatives is best. How is this done?

To test any hypothesis, one must first determine what would happen if the hypothesis is right. In other words, if the hypothesis is correct, then what would you expect to happen under certain conditions? Testing all hypotheses requires thinking that takes this _If-and-then_ form. The _If-and-then_ thinking used to determine correctness of the present hypotheses looks like this:

Hypothesis:	*If . . .* the sunlight coming through the window was awakening the child.
Experiment:	*and . . .* the sunlight is blocked with a heavy cover over the window,
Expected Result:	*then . . .* he will awaken later.

On the other hand,

Hypothesis:	*If . . .* his hunger was awakening him
Experiment:	*and . . .* he is fed an additional bottle of milk at midnight,
Expected Result	*then . . .* he will awaken later.

The result of this type of thinking is the expected result — a *prediction*. An expected result is what you would predict to find if you assume that the hypothesis is correct and conduct the experiment.

What remains to be done to test the hypothesis is to compare the expected result with what in fact happens when you try the experiment (the actual result). If what is expected actually happens, then you have supported your hypothesis. (You place a heavy cover over the window and the child awakens later.) If what happens is different from the expected result, then the hypothesis has not been supported. (The child is fed an additional bottle of milk at midnight, but he still awakens at 5:00 A.M.) If what is expected does not happen, then you must conclude that either something was wrong with your hypothesis or something was wrong with the way you did your experiment. This last phase of trying to answer a question is sometimes called the test phase because the purpose is to test (and so to support or reject) the hypotheses that have been advanced.

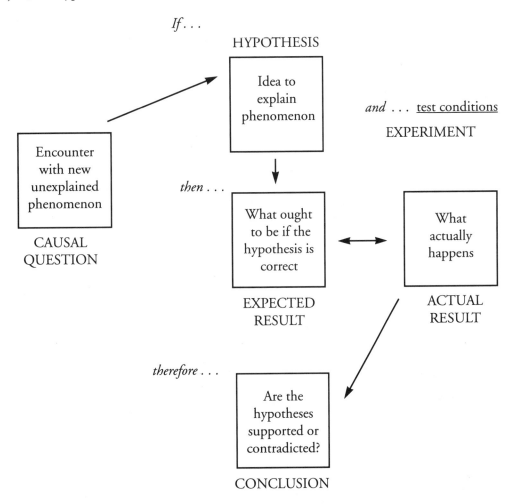

Figure 1. The form of scientific investigation and key words *If . . and . . then . . therefore* used to get from one place to the next.

The pattern of the investigative process is shown in Figure 1.

The first box in Figure 1 represents the first phase — the encounter with some event that raises a question. The arrow represents the creative generation of a hypothesis or hypotheses to tentatively answer the question. The second box represents the invented hypothesis. The third box represents the experimental conditions. The arrow from this box represents logical deduction that results in an expected result of the experiment. The expected result is represented by the third box. The double-headed arrow represents the comparison of the expected result with the actual result of the test (fourth box), resulting in a conclusion of either support or rejection of the hypothesis (last box).

In the present example, the motivation to engage in the investigative process was to solve a rather practical problem: How can we get the early riser to sleep longer? Because the question was one of practical importance and the solution was to be applied directly to solve a practical problem, this research would be termed *applied research*. Some scientists are engaged in applied research. Many scientists, however, are engaged in research trying to answer questions simply for the sake of furthering their understanding. They are simply curious, with no consideration of whether and when their answers may someday be applied for practical purposes. Their research is called *pure research*.

The research in which we were engaged in Investigation 1 — trying to find out why the candle went out and why the water rose — is an example of pure research. Curiosity motivates pure research. Figure 2 shows the phases of the investigative process involved in testing one hypothesis from Investigation 1.

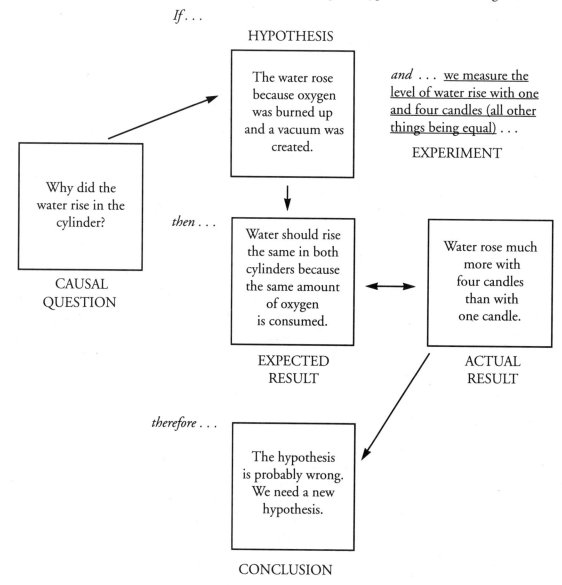

Figure 2. Testing a hypothesis from Investigation 1.

Study Questions

1. Identify a causal question raised in a previous investigation done in class (other than Investigation 1).

2. What alternative hypotheses were generated?

3. What experiments were conducted to test the hypothesis?

4. What expected results were deduced from the hypotheses?

5. Were the results consistent with the expectations?

6. What conclusions were drawn?

7. What is the difference between pure and applied science?

8. Imagine you take a walk in a park and see these two trees with the grass below them.

A   B

a. What question(s) is (are) raised?

_____

_____

_____

b. Generate four alternative hypotheses to answer this question.

_____

_____

_____

_____

_____

c. Describe an experiment to test one of your hypotheses.

_____

_____

_____

_____

d. State an expected result based on the assumed truth of this hypothesis and your experimental design.

_____

_____

_____

Classify sentences 9 – 16, based on the story about homing pigeons in Chapter 8 of the textbook, into one or more of the following sentence types:

a. causal question
b. noncausal question
c. hypothesis
d. experiment
e. prediction (based on deduction)

f. prediction (based on extrapolation)
g. experimental result
h. irrelevant fact
i. conclusion

9. _____ Most homing pigeons are gray.

10. _____ Barney is usually the first pigeon to make it home, so I predict that he will be the first one to return home tomorrow.

11. _____ Frosted contact lenses were placed over the eyes of a number of pigeons.

12. _____ Pigeons wearing frosted contact lenses will not be able to find their way home.

13. _____ Many of the pigeons wearing the frosted lenses were able to find their way home.

14. _____ Homing pigeons use sight to recognize landmarks and get back home.

15. _____ How do homing pigeons manage to find their way home?

16. _____ How long did it take the pigeons to return?

17. Which, if any, of the sentences in 9 – 16 could be classified as more than one type? Give an example of some occasions when someone stated a hypothesis as though it were a conclusion. Explain how this error can lead to misunderstandings and/or inappropriate behavior.

_____

_____

_____

_____

_____

_____

Brain Bender 1

For many students, the process of generating and testing hypotheses seems a bit odd. Instead of simply observing the world to see whether your explanation is correct, one must start by assuming that the explanation is correct so that it may be shown to be incorrect! If the explanation is not shown to be incorrect, then it can be retained for the time being at least. Although this approach to learning about the world may seem backward, it nevertheless is the basic pattern of adult thinking. Try your mind at these two puzzles to see whether you can identify this "backward" thinking pattern.

The Four Index Cards
Below are pictures of four index cards.

| E | K | 4 | 7 |

Each of the cards has a letter on one side and a number on the other side. Read the following rule:

If a card has a vowel on one side, then it has an even number on the other side.

Suppose you want to test whether this idea is correct or incorrect for these four cards. Which of the four cards must be turned over to allow the rule to be tested? Explain.

_____

_____

_____

_____

Brain Bender 2

The Islands Puzzle

The puzzle is about four islands in the ocean — Bean Island, Bird Island, Fish Island, and Snail Island. People have been traveling among these islands by boat for many years, but recently an airline started in business. Carefully read the clues given below about possible plane trips. The trips may be direct, or they may include a stop on one of the islands. When a trip is possible, it can be made in both directions between the islands.

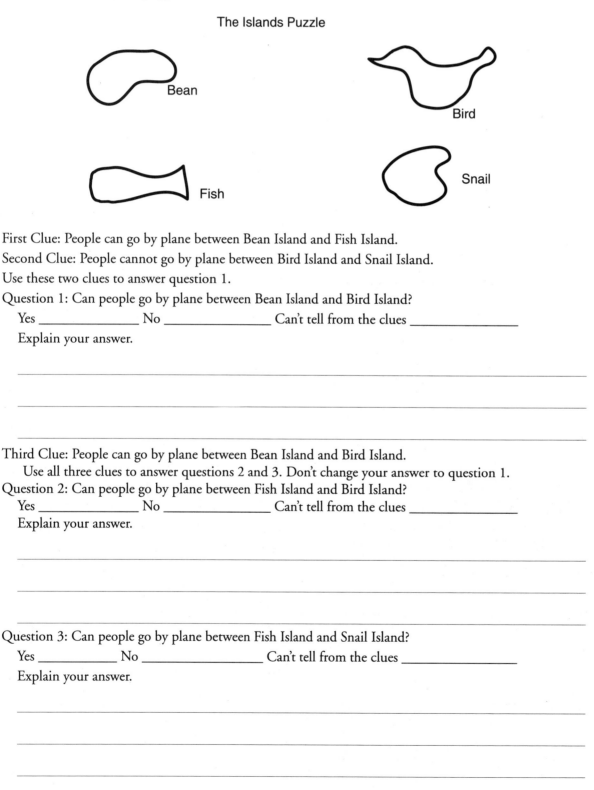

The Islands Puzzle

First Clue: People can go by plane between Bean Island and Fish Island.

Second Clue: People cannot go by plane between Bird Island and Snail Island.

Use these two clues to answer question 1.

Question 1: Can people go by plane between Bean Island and Bird Island?

Yes _____ No _____ Can't tell from the clues _____

Explain your answer.

_____

_____

Third Clue: People can go by plane between Bean Island and Bird Island.

Use all three clues to answer questions 2 and 3. Don't change your answer to question 1.

Question 2: Can people go by plane between Fish Island and Bird Island?

Yes _____ No _____ Can't tell from the clues _____

Explain your answer.

_____

_____

Question 3: Can people go by plane between Fish Island and Snail Island?

Yes _____ No _____ Can't tell from the clues _____

Explain your answer.

_____

_____

Investigation 14 *What Happens During the Development of Chicken Eggs?*

Introduction

Few people realize that a popular source of breakfast protein is actually a large cell — the chicken egg. You are familiar with the structures and functions of some microscopic cells. Other than size, is a chicken egg like the other cells you have observed? If not, how does it differ? What structures can be found in a chicken egg? What are they for? What happens to them during the development of a new chicken? In this activity, you will find out.

Objectives

1. To identify structures of an unincubated chicken egg and create alternative hypotheses about possible functions.

2. To observe a series of incubated eggs for evidence suggesting functions for cell structures.

3. To observe and record the development of a chick embryo.

4. To list the basic processes that occur during embryonic development of a chicken and compare these to development in other organisms.

Materials

paper towels
culture dish
chicken egg, unincubated
dropper
forceps
dissecting microscope
petri dish
scissors
0.9% sodium chloride solution (saline)
70% alcohol (ethanol)
chicken eggs incubated to 48 and 72 hours
  and to 4, 5, and 10 days

Procedure Day : Unincubated Egg

1. The central questions we are trying to answer are these: *What structures are in the chicken egg? What happens to them during development?* To begin to answer these questions, crumple a paper towel in the bottom of a culture dish. Place the unincubated egg on its side in the dish so that it cannot roll. **Caution: Wear your laboratory apron.**

2. Remove the upper part of the shell by gently piercing the shell with the scissors and very carefully cutting away an oval top. See the diagrams. With a dropper, draw off the fluid until the yolk is no longer covered. Carefully remove more of the fluid and shell until only about half of the eggshell remains. **Caution: Do not taste or eat any part of the chicken eggs. Wash your hands thoroughly after handling eggs.**

3. Observe the contents of the egg. What major subdivisions can be seen? The central portion (the egg yellow) is called the *yolk.* What are some possible functions of the yolk?

_____

_____

_____

4. What may happen to the yolk during development?

_____

_____

_____

5. Try to locate a tiny whitish spot on the yolk. It may not be visible. If you cannot see it, coax the yolk around by tugging gently at the twisted cords or by pushing the yolk gently with the forceps. Use a dissecting microscope to observe this structure. Do you see more than one white spot? Can you measure the spot? What are some possible functions?

_____

_____

_____

6. Now carefully put all the contents of the egg, except the shell, into the culture dish that contains the paper towel. Examine the shell. Write a description of what you see.

_____

_____

_____

7. Draw up some of the fluid portion surrounding the yolk into a dropper. Place a few drops on your fingers. Write a description of the fluid. What are some possible functions?

_____

_____

_____

8. Locate the whitish, twisted cords close to the yolk. Carefully grasp one of these cords and pull gently. Does the cord appear to be attached to the yolk? What do you suspect these are for?

_____

_____

_____

9. Draw a diagram of the egg. Use arrows to point to each structure you have observed. Beside each arrow, write alternative hypotheses for the function of each structure.

10. Assume that your hypotheses are correct. What would you expect (predict) the observed structures to look like after the egg has started to develop? Sketch and be prepared to discuss your ideas during a class discussion.

11. Does the yolk appear to be contained by a membrane? See whether you can find out. What may be the function of the membrane?

_____

_____

_____

12. Construct and complete a table like Table 1 during the course of your observations of the egg series.

Table 1 Egg Structures and Functions

Egg Structure	Hypothesized Function(s)	Expected Change	Actual Change

Procedure Day 2: 48-Hour Egg

1. Obtain a 48-hour incubated egg from your teacher.

2. Open the egg the same way you opened the unincubated egg.

3. Observe this egg with a dissecting microscope and compare it with your original drawing. Make a careful drawing of the egg, showing as much detail as possible. Below your drawing, list similarities and differences between this egg and the first one you observed.

4. Keep records describing developmental events as they occur on a chick development table (see Table 2).

Procedure Day 3: 72-Hour Egg

1. Open the 72-hour egg as follows. Support the egg, large end up, on a crumpled paper towel in a petri dish. Very gently crack the large end of the shell with the scissors handle. Use the forceps to pick away the shell. Avoid breaking the shell membrane if possible. After part of the shell has been removed (about one-third), put the egg into a culture dish of saline (0.9% sodium chloride) solution (at 38°C) and pick off the remainder of the shell. Now remove the dense white membrane covering the embryo. Write a description of what you see.

_____

_____

2. Make a careful drawing and list similarities and differences between this egg and the 48-hour egg. Record in Table 1 any additional information about function of structures. Record developmental events in Table 2.

3. Compare your observations with those of others during a class discussion.

Procedure Day 4

1. Open the 4-day egg the same way you opened the 3-day egg. Write a description of what you see.

_____

_____

2. Make a careful drawing. Record appropriate information in Tables 1 and 2.

3. When you have completed your observations, place the embryo in a preservative solution (70% alcohol) provided by your teacher. **Caution: Alcohol is poisonous. Be careful not to inhale fumes. Wear your safety goggles.**

4. Be prepared to discuss your observations.

Procedure Day 5

1. Obtain a 5-day incubated egg and open it as you did the 4-day egg.

2. Make a careful drawing and record observations on your tables.

3. Remove the membrane covering the embryo and locate the yolk. Does your observation of the yolk suggest a function for it? Explain.

_____

_____

4. Be prepared to discuss your observations.

Procedure Day 10

1. Obtain and open a 10-day incubated egg as you did the 5-day egg.

2. Make your drawing and record your observations.

3. Be prepared to discuss your observations.

Table 2
Chick Development

Structure	48 Hours	72 Hours	4 Days	5 Days	10 Days
Head Development					
Beak					
Ear Opening					
Eyelids					
Neck					
Egg Tooth					
Limb Development					
Wings/Feathers					
Legs/Scales					
Toes					
Claws					
Body Feathers					
Heart Development					
Tubular					
Twisted					
Enclosed in body cavity					
Beating					

Study Questions

1. Which of the alternative hypotheses proposed on Day 1 were supported? Which were contradicted? Explain.

2. In light of your observations, what organ seems to be the first to develop in the chick?

3. What appears to be the purpose of the egg tooth?

4. What is cell differentiation? Why is cell differentiation important?

5. What may be the relationship between genes and the process of differentiation?

6. If an ice pack is kept on the back of a Himalayan rabbit for a period of time, the hair under the pack will grow in black, gradually replacing the white hair. What does this suggest about the influence of environment on gene expression?

7. The eggshell contains a large amount of calcium. At Day 8 of incubation, the shell calcium level begins to decrease. After Day 14, the calcium level in the shell decreases even more drastically. Use your observations of the developmental events of the embryo to develop alternative hypotheses about why the calcium level of the eggshell decreases.

8. The surface of the eggshell is pierced with fine pores (0.04-0.05 mm diameter). Each shell has approximately 7,000 pores. Write one or more hypotheses to explain the function of these pores and design an experiment to test one of them. Be sure to state expected results and to include controls in your experimental design.

_____

_____

_____

_____

_____

_____

_____

_____

9. What major similarities and differences exist between chicken and human development?

_____

_____

_____

_____

Investigation 15 *What Are Foods and Beverages Made Of?*

Introduction

Everyone knows that we must eat and drink to stay alive and to grow. You no doubt have been told to eat a "balanced" diet — that you must eat a variety of foods and beverages to obtain the multitude of molecules needed by your body. But how do we know which foods and beverages to eat? Do procedures exist for discovering what molecules various foods and beverages contain? The answer is yes. In this activity, you will have an opportunity to test your ideas about the molecules contained in a variety of familiar foods and beverages. You will use a number of chemical indicators to do this.

Objectives

1. To use procedures for testing foods and beverages for the presence or absence of specific types of molecules.

2. To group common foods and beverages based on their reactions to indicators.

3. To invent names for specific types of molecules found in common foods and beverages.

Materials

foods and beverages	test-tube racks
mortar and pestle	glass stirring rods
distilled water	indicator materials:
safety goggles	iodine
hot plate	Benedict's solution
test-tube clamps	Biuret reagent
test tubes	copper sulfate solution
wax pencils	sodium hydroxide
hot-water bath (250-mL beaker)	

Procedures

1. Your teacher will provide a list of common foods and beverages. List them in the left-hand column of your data table (see example). The specific question we hope to answer is, *What types of molecules are these foods and beverages made of?*

2. You may already know something about what these foods and beverages are made of. With your group members, attempt to group the foods and beverages into four to nine groups based on what types of molecules you think they may contain. Record your grouping.

_____

_____

3. Now systematically subject each food and beverage to a specific set of procedures (see step 4) to see how it reacts. Keep a record of those reactions. Because the reactions will depend on what the foods and beverages are made of, they will provide you with a way of seeing how accurate your initial groupings were. In other words, if your initial groupings are correct, then all of the foods and/or beverages in a specific group should react to the indicators in the same way. Record the reactions in your table. If a color change occurs, write what the change is. If no change occurs, write *NEG* (for negative) in the appropriate space.

4.   If your sample is a solid, crush a pea-sized amount with the mortar and pestle and add a small amount of distilled water to it before testing. **Caution: Wear safety goggles during the tests. Also use care in handling the Biuret reagent; it may burn skin or clothing. Sodium hydroxide is very caustic.**

- Iodine Test — Add a few drops of the iodine solution to the unknown solution.
- Benedict's Solution — Add 5 mL of Benedict's solution to 8 drops of the unknown solution in a test tube. Shake well. Place the test tube in a boiling water bath for 3 minutes. Set the solution aside to cool. When cool, check for a color change.
- Paper Test — Drop or rub a small amount of the unknown on a piece of indicator paper. Check for the appearance of a translucent spot. If no translucent spot appears, the test is negative (NEG).
- Biuret Test — Mix 2-3 mL of the unknown solution with an equal amount of sodium hydroxide. Then add, drop by drop, a 0.5% copper sulfate solution. Use a stirring rod to swirl the contents of the tube gently between drops.
- Limewater — Add drops of the unknown solution to a test tube partially filled with limewater (optional).
- Bromthymol Blue (BTB) — Add drops of the unknown solution to a test tube partially filled with a dilute BTB solution.

  or

- Phenolphthalein — Add drops of the unknown solution to a test tube partially filled with phenolphthalein solution.

Data Table
Results

Food or Beverage	Iodine	Benedict's	Paper	Biuret	Limewater	BTM	Phenolphthalein
1.							
2.							
3.							
4.							
5.							
6.							
7.							
8.							
9.							
10.							
11.							
12.							
13.							
14.							
15.							

5. Compare your results with your initial groupings. Did all the foods/beverages that you had grouped together react in the same way? If so, what type of molecules do you suspect members of each group contain that are causing the similar reaction? If not, do you have any ideas for a new grouping system? What is the basis of that system? Be prepared to discuss your ideas in a class discussion.

_____

_____

_____

_____

6. Have one member of your group record your data on the class data sheet. How do your results compare with those of other groups? How can differences be explained?

_____

_____

_____

Study Questions

1. In a class investigation, five drops of iodine indicator were added to five mL of a clear food substance. The resulting solution was light brown in color. What most likely happened to the molecules in the test tube?

_____

_____

_____

2. What kinds of molecules are most likely indicated by each of the indicators?

_____

_____

_____

3. Are foods and beverages made up of more than one kind of molecule? Which ones? What evidence do you have for your answer?

_____

_____

_____

4. Draw the structure of a simple protein, starch, sugar, and fat molecule. What evidence do biochemists have that suggests these structures?

_____

_____

_____

5. How does a chemical reaction differ from simply mixing molecules together? How can one tell when a chemical reaction has probably occurred?

_____

_____

_____

6. In the table below, a + indicates a color change. A blank indicates no change.

Indicator

Food or Beverage	Iodine	Benedict's	Paper	Biuret	BTM	Phenolphthalein
1.	+					
2.		+				
3.					blue	red
4.						
5.					yellow	
6.			+			
7.		+		+	blue	

State whether each of the following statements is *true, false,* or *not possible to determine from the available data.*

a. Unknown 1 contains chains of sugar molecules.

_____

b. Unknown 2 contains simple sugar molecules (not in chains).

_____

c. Unknown 3 has a relatively high number of hydrogen ions.

_____

d. Unknown 4 may be distilled water.

_____

e. Unknown 5 may be distilled water.

_____

f. Unknown 6 has a pH less than 7.

_____

g. Unknown 7 contains protein but no sugar and has a pH greater than 6.

_____

Investigation 16 *What Is the Function of Saliva?*

Introduction

Have you ever eaten a dozen crackers without anything to drink? Gets dry, doesn't it? But if you wait a few minutes, your mouth becomes moist again and you could eat a thirteenth cracker if you had to, still without anything to drink. Why? Somewhere in your mouth you produce saliva, and saliva moistens food and makes it easy to swallow. Is that all saliva does? As you may know, certain liquids produced in the stomach and intestine actually break down food molecules into smaller ones so that they can enter the bloodstream. Perhaps saliva also helps start the breakdown process. What else might it do? In this investigation, you will design and conduct experiments to test the various hypotheses that are advanced.

Objective

To design and conduct controlled experiments to test alternative hypotheses about the function(s) of saliva.

Materials

test tubes

test-tube rack

spot plates

stirring rods

mortar and pestle

hot-water bath

Benedict's solution

iodine solution

Biuret reagent

saliva (or "substitute saliva" solution)

food samples:

potato

cracker

honey

artificial sweetener

lunch meat

corn syrup

Procedure

1. List possible answers (alternative hypotheses) to the following question: *Other than moistening and lubrication, what might be some other functions of saliva?* Be prepared to share your ideas in a class discussion.

_____

_____

_____

2. Select what you believe to be the best of the alternative hypotheses advanced and design an experiment or experiments to test it. Recall that previous work has shown that the presence or absence of certain molecules in foods can be discovered by adding specific indicators to the food and observing their reactions. Your teacher will review procedures used for each of the following indicators: Biuret, iodine solution, Benedict's, paper. The indicators indicate the presence of protein, starch, glucose, and fat, respectively, as shown below. **Caution: Wear safety goggles during the tests. Use care in handling the indicators. Biuret reagent may burn skin and clothing. Sodium hydroxide is caustic. Use caution when handling glassware.**

Indicators and Reactions

Indicator	Tests For	Reaction
Biuret	Protein	Changes color from pink to purple
Iodine	Starch	Changes color from orange to blue-black
Benedict's	Glucose	Changes color from blue to green to yellow to orange
Paper	Fat	Translucent, greasy spot on paper (compare it to a water spot)

3. Before starting your experiments, think ahead to their completion. Assume your hypotheses are correct. State your expected results (what do you think will happen?).

_____

_____

_____

4. Conduct your experiments and record results in a data table of your design.

5. 〔☠〕 If for some reason you do not wish (or are not allowed) to use your own saliva for this experiment, ask your teacher for a prepared saliva solution that contains molecules normally found in saliva. If you use your own saliva, your teacher can show you the best way to collect it. **Caution: Do not eat food in the laboratory as part of your experiment.**

Study Questions

1. Did your expected results stated in step 3 match what actually happened? If not, which should be changed — your hypothesis, or your experimental design? Explain.

_____

_____

2. In light of the results of this lab, what can you conclude about the function of saliva?

_____

_____

3. Did you conduct controlled experiments? Explain.

_____

_____

4. What were the dependent and independent variables in your experiments? Which variables should have been held constant?

_____

_____

5. How does chemical breakdown differ from mechanical breakdown? How are the processes similar?

_____

_____

6. Obtain a copy of the following diagram in the back of this lab manual. Fill in the boxes and blanks with one causal question, hypothesis, experiment, expected results, actual results, and conclusion from this investigation. (See Reasoning Module 5 for details.)

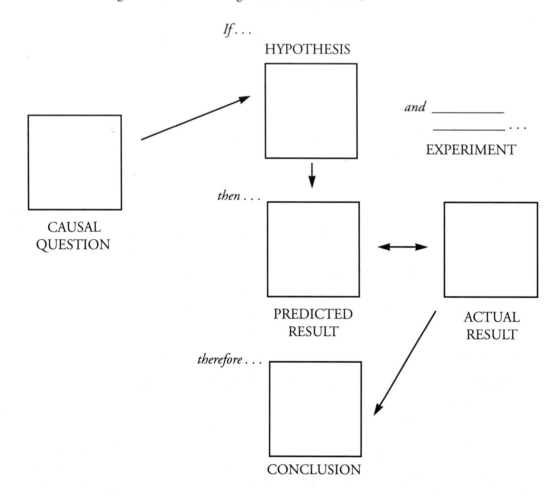

Investigation 17 *What Happens to Molecules During Chemical Breakdown?*

Introduction

A common household chemical consists of hydrogen peroxide molecules (H_2O_2). You can pour it on cuts, in ears, on hair, wherever. What would happen if you drank it (DO NOT try to find out!) — or fed it to your dog (don't try this either)? What would it do to your — or Fido's — internal organs? One animal organ that is easy to obtain is beef liver. Conduct the following test to determine what happens when hydrogen peroxide molecules come in contact with beef liver. What happens to the liver? To the hydrogen peroxide? To your hand?

Objective

To test alternative hypotheses concerning the interaction of liver and hydrogen peroxide molecules.

Materials

raw liver
hydrogen peroxide (H_2O_2)
dropper
graduated cylinder
test tubes
test-tube rack
glass stirring rods
scissors
forceps

Procedure

1. Put 20 drops of hydrogen peroxide into a small test tube. Hold the tube and use forceps to add a small piece of liver about the size of a bean seed. **Caution: Hydrogen peroxide is poisonous. Do not taste or eat any of the liver. Caution: Use care when handling glassware.**

2. When you added the liver to the hydrogen peroxide, what happened to the liver? To the hydrogen peroxide? To your hand?

_____

_____

3. The key question is this: *What might be happening to the molecules in the hydrogen peroxide and in the liver?* The following are three possible accounts of what is happening to the molecules in the liver and in the hydrogen peroxide. List any other alternative hypotheses that you can think of. Be prepared to discuss these and your ideas in class.

   a. The molecules in the liver and the hydrogen peroxide both changed to form a new substance (new types of molecules).

   b. The hydrogen peroxide molecules changed, but the liver molecules did not.

   c. The liver molecules changed, but the hydrogen peroxide molecules did not.

4. Which of the hypotheses listed in step 3 do you suspect might be the best explanation? Explain.

5. Try to think of an experimental design that could determine which explanation(s) is (are) correct. What are the expected (predicted) results of your hypotheses and experiments? Be prepared to compare your ideas with those of others in a class discussion. Record your ideas.

6. [skull icon] Carry out your experiment(s). Record your design and results. **Caution: Be sure to dispose of the liver properly when you have completed your experiment, and to wash your hands thoroughly.**

Study Questions

1. State the alternative hypotheses tested by your design. State the expected results generated from your hypotheses and experimental designs.

2. Do the data support the hypotheses proposed? Explain.

3. Has your experiment allowed you to eliminate other possible explanations? Explain.

4. What might have caused the temperature to increase during the reaction?

5. Did you conduct a controlled experiment? Explain.

6. What were the independent and dependent variables in your experiment?

_____

_____

7. Define the terms *catalyst* and *enzyme*. How do enzymatic reactions differ from nonenzymatic chemical reactions?

_____

_____

8. How do chemical reactions differ from simple mixing of molecules?

_____

_____

9. Obtain a copy of the following diagram in the back of this lab manual. Fill in the boxes and blanks with one causal question, hypothesis, experiment, expected result, actual result, and conclusion from this investigation. (See Reasoning Module 5 for details).

_____

_____

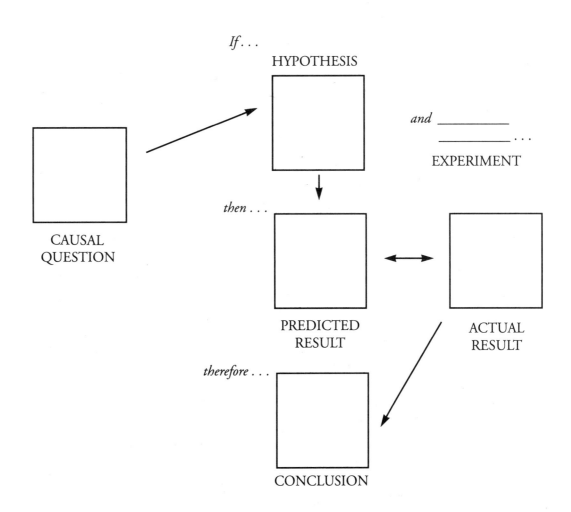

Investigation 18 *How Do Molecules Pass Into and Out of Cells?*

Introduction

A single-celled organism is surrounded by a cell membrane that serves as a barrier between the cell's interior and the external environment. For the organism to survive, however, food and water molecules must pass through the cell membrane into the interior, and waste molecules must pass out. The same situation holds in a multicellular organism. Your digestive system is designed primarily to break up food into molecules small enough to pass into your body cells through their cell membranes. Likewise, waste molecules must be able to leave cells to be carried away by the circulatory system. But how do molecules actually pass through cell membranes?

In this investigation, you will test the effect of two independent variables (molecular size, and concentration of molecules) on the passage of molecules through a cell membrane and attempt to create a theory of membrane transport.

Objectives

1. To investigate variables that influence the passage of molecules into and out of cells.

2. To create a theory to explain the passage of molecules through cell membranes.

Materials

slides and coverslips
elodea cells (a freshwater plant)
compound microscope
5% starch suspension:
  5 g corn starch dissolved in 95 mL distilled water
red blood cells (from an animal such as a cow)
droppers
dialysis tubing
string
250-mL beakers
balance
paper towels
iodine solution
Benedict's solution
1% starch suspension:
  1 g corn starch dissolved in 99 mL distilled water
1% glucose solution:
  1 g glucose dissolved in 99 mL of distilled water
5% glucose solution:
  5 g glucose dissolved in 95 mL of distilled water
  100% water solution (100 mL of distilled water)

Procedure Part I — Elodea Cells and Red Blood Cells

1. Work with one or two lab partners to do this lab. Remember the central question you are trying to answer is: *How do molecules pass in and out of cells?* Start by preparing a slide of elodea, using water from the container of elodea. Make sure to use a coverslip. **Caution: Be careful when handling slides and microscope.**

2. Find several cells under high power and make sure you can identify individual cells, their relatively thick cell walls, and their chloroplasts (see diagram). The thin cell membranes line the inside of the cell walls but will not be visible.

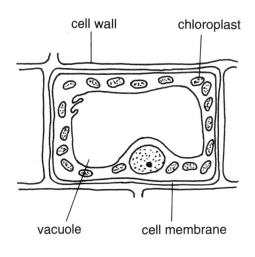

3. What happens to an elodea cell when it is placed into a 5% starch suspension? What happens when an elodea cell is placed into distilled water? To find out, prepare one slide of an elodea leaf in 5% starch suspension and another slide in distilled water. On each slide, locate a few cells under high power. Observe the location of the chloroplasts in relation to the cell wall. Compare the appearance of the cells in each of the two solutions to those in step 2. Sketch your observations.

4. Propose at least two alternative hypotheses (explanations) for your observations.

_____

_____

_____

5. ☠ What happens to red blood cells when placed in 5% starch suspension? What happens to red blood cells when placed in distilled water? To find out, prepare a slide of red blood cells in 5% starch suspension and another in distilled water. To do this, add a drop of blood to a slide and cover it with a coverslip. Use a dropper to add a drop of suspension to the blood while you are observing it under high power. Observe cell appearance and repeat with the distilled water. Sketch your observations. **Caution: Do not put any animal tissue in your mouth.**

6. Propose at least two alternative hypotheses to explain your observations in step 5. Be prepared to share your ideas in a class discussion. Part II of this investigation will be devoted to testing some of your ideas by using "model" cells.

_____

_____

_____

Procedure Part II — Experimenting With Model Cells

1. A water molecule (H_2O) is made up of two hydrogen atoms and one oxygen atom as shown below. A sugar molecule made up of 6 carbon atoms, 12 hydrogen atoms, and 6 oxygen atoms ($C_6H_{12}O_6$), also shown on page 96, is called glucose. When glucose molecules join with other glucose molecules, they form long chains that loop around. These long, looping chain molecules, such as the one shown below, are called starches.

   Notice that the materials list includes distilled water (water molecules only), as well as starch molecules mixed in suspension with water molecules in varying concentrations. A single molecule of glucose is about ten times the size of a single molecule of water; thus, a single molecule of starch is many, many times larger than a molecule of water. Do your observations from Part I suggest that the size of a molecule and/or its concentration might affect its movement through a cell membrane? If so, why? Record your ideas.

CH₂OH

water   glucose   starch

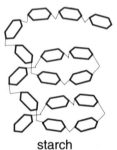

2. 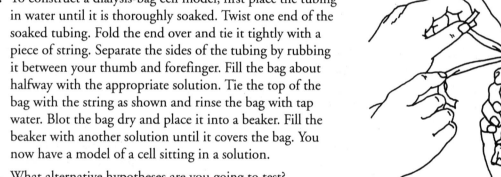 Design at least three experiments to test your ideas, using the materials provided. You may wish to discuss your ideas with your teacher or classmates before you proceed. In any event, using the dialysis tubing and string, you will need to construct at least three model cells. Keep in mind that the internal and external solutions can be varied. You can detect the movement of substances into or out of your model cells by finding their weights (masses) at various time intervals. Plan to graph your results comparing model cell mass (in grams) on the vertical axis with time (in minutes) plotted on the horizontal axis. When starch suspensions mix with iodine, they turn purple; thus, you may be able to use this color change as an indication of movement. Benedict's solution can be used to detect glucose. **Caution: Iodine stains clothing and skin. Wear your safety goggles and laboratory apron.**

3. To construct a dialysis-bag cell model, first place the tubing in water until it is thoroughly soaked. Twist one end of the soaked tubing. Fold the end over and tie it tightly with a piece of string. Separate the sides of the tubing by rubbing it between your thumb and forefinger. Fill the bag about halfway with the appropriate solution. Tie the top of the bag with the string as shown and rinse the bag with tap water. Blot the bag dry and place it into a beaker. Fill the beaker with another solution until it covers the bag. You now have a model of a cell sitting in a solution.

4. What alternative hypotheses are you going to test?

_____

_____

5. Briefly describe your experimental procedures (in enough detail that someone else could repeat your experiments).

_____

_____

_____

6. What are your expected (predicted) results?

_____

_____

7. What are the independent and dependent variables in your experiments?

_____

_____

8. Conduct your experiments. Record and graph your results. Be prepared to graph your results on the board for a class discussion.

9. Following the class discussion, summarize your results and those of others. Did a consistent pattern emerge from the data? If so, what was it?

_____

_____

10. What do the results suggest about the acceptability of the various hypotheses advanced to explain what happened to the elodea cells in Part I? Explain.

_____

_____

11. Describe the appearance of a hypothetical cell membrane that is consistent with your experimental observations from this lab. What is the best answer your data allow you to give to the question, *What causes molecules to pass into and out of cells?*

_____

_____

_____

Study Questions

1. Given what you know about diffusion, would you expect lightweight or heavy molecules to move faster? Explain.

_____

_____

2. Drinking salt water will make a person thirsty. If a person continues to drink salt water, he or she will die of thirst. Explain why a person gets thirsty when drinking salt water.

_____

_____

3. What effect do you think a rise in temperature would have on diffusion rate? Explain.

_____

_____

4. When red blood cells are placed in distilled water (pure water), they will

   a. lose water and shrink.

   b. gain water and burst.

   c. stay the same.

   Explain your choice.

_____

_____

5. A student set up an experiment to see whether dextrose, a complex sugar, moves through a cell membrane. The student placed exactly 50 mL of distilled water into a dialysis tubing cell and 300 mL of a 5% dextrose solution outside of the cell.

   a. What will probably happen to the volume of liquid inside the cell after several hours. Explain.

   _____

   _____

   b. Is the concentration of dextrose likely to increase or decrease inside the cell? Explain.

   _____

   _____

6. What were the independent variables tested in your investigation using model cells? What were the dependent variables? Name three variables that should have been held constant.

   _____

   _____

   _____

7. State one hypothesis you tested, and state one expected result that could be logically derived from the hypothesis and your experimental design. Was the hypothesis supported or not? Explain.

   _____

   _____

   _____

   _____

8. Suppose a molecule moves into a cell through a hole in the cell membrane. What force or forces might have caused it to move in that direction?

   _____

   _____

   _____

9. Molecules tend to move from an area of high concentration to an area of low concentration. Why?

   _____

   _____

   _____

10. Obtain an enlarged version of the diagram in the back of this lab manual. Fill in the boxes and blanks with one causal question, hypothesis, experiment, expected result, actual result, and conclusion from this investigation. (See Reasoning Module 5 for details.)

Investigation 19 *What Is the Molecular Structure of Genetic Material?*

Introduction

One of the most fascinating stories in the history of biology is the creation and test of a model of the molecular structure of the genetic material contained in chromosomes. In this investigation, you will be given the opportunity to become familiar with a number of important discoveries about the chemical nature of chromosomes, and you will attempt to create and test a model of the molecule that carries genetic information. The model should tell you how information is passed from parent to offspring. Creating such a model will not win you a Nobel Prize, but it did win one for the American James Watson and the Englishman Francis Crick in 1962.

Objectives

1. To use historical clues to construct a functional molecular model of the genetic material contained in chromosomes.

2. To better understand how theory and data interact to generate new knowledge.

Materials

historical clues about chromosome structure
scissors
crayons and/or colored markers
sheets of molecule cutouts
cellophane tape

Procedure

1. Work with one or two other students to do this activity. The central question you are trying to answer is, *What is the molecular structure of the genetic material contained in chromosomes?* To answer this question, you will do essentially what Watson and Crick did. They used a set of molecular models and the available chemical and physical data about chromosomes to create and test alternative hypotheses about molecular structure — and so will you.

2. To start, obtain the first set of historical clues from your teacher. Read the six clues.

3. If it is assumed that chromosomes carry the genetic information, the following question is raised: *What molecule in the chromosomes carries the genetic information?*

_____

_____

4. What two alternative hypotheses are suggested by the clues?

_____

_____

5. Which of these alternative hypotheses seems most likely to be able to be supported? Why?

_____

_____

6. Using the first set of clues, attempt to construct a model of the DNA molecule. To do this, you will need first to cut out the molecular models found on pages provided by your teacher. You may find it is helpful to color code the nucleotide base models by outlining their borders as follows:

adenine	— red	guanine	— green
cytosine	— blue	thymine	— yellow

7. After you have cut out at least 20 of each of the model molecules, see whether you can arrange them to make a DNA molecule. At this point, about how many different combinations are consistent with the clues? Be prepared to discuss the possibilities in a class discussion.

8. After the class discussion, obtain the second set of clues and refine your model in light of this new evidence. Be prepared to share your ideas in a class discussion. What questions remain?

9. Obtain the third and final set of clues and try to arrive at a single configuration that is consistent with all of the evidence. Be prepared to present the model and your arguments to the class.

Study Questions

1. What type of nucleotides must be attached to each other to make your model stable? Can you be sure your model is correct? Why or why not?

2. What percentage of your model nucleotides is adenine, cytosine, guanine, and thymine? Show your work. How do your percentages compare with those presented in the table of percentages obtained biochemically (from clue set 3)?

3. The information in question 2 was very important in the creation of the Watson-Crick model of DNA. What does it tell you about your model? Does the information from questions 1 and 2 further strengthen your hypothesized model?

4. X-rays are known to break apart the bonds that hold the DNA molecule together. Doctors seldom X-ray patients who are in early pregnancy. Why?

5. Genetic engineering involves changing DNA molecules. How might genetic engineering be used in the battle against "germs" and disease?

6. You have studied mitosis and genetics. What do you think the DNA molecule must do in order for the correct message to be sent to the next generation? With your group, try to formulate a way in which DNA might accomplish this. Remember that in order for the correct information to be sent to the daughter cells, replication must be exact. Use the cutout models to simulate DNA replication.

Investigation 20 *Where Will Brine Shrimp Eggs Hatch?*

Introduction

Animals, like plants, are affected by favorable and unfavorable factors in their environment. Unlike most plants, however, adult animals generally are able to avoid unfavorable environmental factors by moving. But many animal eggs face the same problems faced by the seeds of plants in that they too are released directly into the environment for dispersal.

Investigation 18 explored the passage of solutions into and out of cells. In this activity, you will investigate the effect of one environmental factor — varying saltwater concentration — on the hatching of a particular type of cell: the brine shrimp egg.

Brine shrimp are small aquatic animals commonly found in bays and estuaries along coastal regions. One factor that varies in the brine shrimp's environment is saltwater concentration. Do you think this factor may influence the hatching of brine shrimp eggs? If so, why?

Objectives

1. To determine by experiment the effect of varying saltwater concentration on the hatching of brine shrimp eggs.

2. To gain experience in solving numerical problems involving proportions.

Materials

hand lens
saltwater solution (3 times the concentration of normal sea water)*
graph paper
wooden splints
brine shrimp eggs
6 vials
small spoons
aged tap water or distilled water
vial stand
*Normal sea water contains approximately 35 g of salt dissolved in every 1000 mL of water.

Procedure

1. Work with one or two partners.

2. Refer to the map on page 103. Brine shrimp are commonly found in the sea and river estuary shown on the map. On occasion, they are even found some distance upriver and in saltwater pools like the one shown on the map. The saltwater pool shown on the map is often isolated from the sea, and it loses a lot of water by evaporation. Occasionally, its salt concentration reaches three times that of the sea.

3. Design and conduct an experiment that will answer these questions:

   a. What concentration of salt water is the most favorable for the hatching of brine shrimp eggs?

   _____

   _____

b. How far up the river shown on the map would brine shrimp eggs hatch? (Answer in terms of concentration of salt water.)

c. How would the high salt concentration in the salt water pool affect the hatching of brine shrimp eggs?

4. Describe your experimental design. What is the independent variable? The dependent variable? What variables should be controlled? Record your data.

5. At the conclusion of your experiment, graph your data on a separate sheet of graph paper. Be prepared to compare your graphs with those of other students.

Study Questions

1. Name three environmental factors that are likely to limit the population of adult brine shrimp. In what ways are these factors limiting? What do you think would be the optimum range for each factor? Why?

2. Is the human population of the city or town you live in growing, shrinking, or holding steady? What factors appear to be responsible for its present condition? If the population is growing, what factors may limit its growth in the future? What population size do you think would be optimum? Why?

3. Suppose you are given a sample of salt water ten times the concentration of normal sea water. How would you dilute this to produce samples of water 1/3 X, 2/3 X, 1 X, 2 X, and 3 X the concentration of normal sea water?

4. Suppose the original water sample in question 3 was seven times the concentration of normal sea water. How would you dilute this sample to produce the concentrations listed?

_____

_____

_____

5. List four independent variables that should be held constant in an experiment to determine the effect of salt concentration on the hatching of brine shrimp eggs. What is the dependent variable in this experiment?

_____

_____

_____

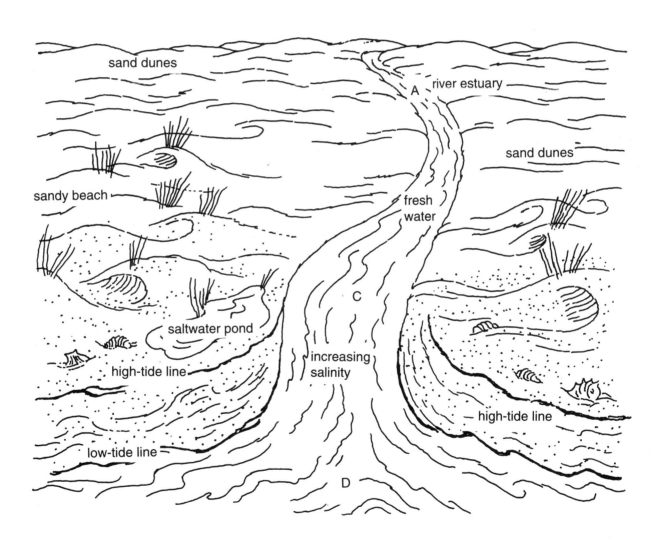

Reasoning Module 6 *Proportional Relationships*

Introduction

If you have ever watched a good golfer, you have probably noticed that the harder he or she swings the club, the farther the ball goes. Things are quite the opposite for most novices, however. The harder they swing, the shorter distance the ball goes. In fact, if the novice swings too hard, he or she may even miss the ball completely! Many other relationships such as these exist in the environment. The farther away an object is, the smaller it appears. The more food people eat, the more weight they gain. The faster a car travels, the more gasoline per mile it uses. The longer a person's legs, the fewer steps he or she must take to get from the TV to the refrigerator. The more a person exercises, the less his or her chances are of dying from heart disease.

Such relationships are called *correlations*. The identification of correlations is a crucial step to understanding the environment; the correlations suggest the possibility of a cause-effect relationship. Not only is it important to be able to identify correlations, it is often important to be able to quantify such relationships. For example, a correlation exists between the amount of something you buy and the price you pay for it. The more you buy, the more you pay. But do you know whether you get a better deal when you buy 2 gallons of ice cream for $1.95 or 5 gallons of the same kind of ice cream for $4.75?

Objective

To solve quantitative problems involving proportional relationships.

Procedure

1. Begin by reading Essay 1. Do the suggested activities as you proceed through the module.

2. Work through some or all of the study questions until you are sure you understand the ideas introduced.

3. For assistance consult a classmate and/or your teacher.

Materials

Cuisenaire® rods of various colors and lengths
balance beam with a set of 10 weights
coupled gear systems

Essay 1: How High Will They Bounce?

Suppose the girl with the white tennis ball had actually conducted a fair test to find out whether her friend's orange tennis ball was bouncier than her white one. This would have been simple to do. She could just have dropped both balls under identical conditions and carefully noted which bounced higher. The one that bounced higher would clearly have been the bouncier ball.

If she had done the test, she would have found out that when the balls are dropped from waist high, the white ball bounces 30 cm into the air and the orange ball bounces 60 cm into the air.

Now suppose the balls are dropped from near her shoulders and the white ball bounces 60 cm into the air. How high will the orange ball bounce under identical circumstances? If you had predicted 120 cm, you would have been correct. It seems that for every 30 cm the white ball rises, the orange ball will rise 60 cm. Therefore, when the white ball rises 60 cm, the orange ball will rise 60 x 2 = 120 cm.

Consider another example of this sort of quantitative relationship before a general method of dealing with such relationships is introduced.

Essay 2: Building Walls

Obtain a set of Cuisenaire® rods and place them on a table. See whether you can construct some "walls" using the Cuisenaire® rods.

Begin by placing one red block in front of you. Now place two white (noncolored) blocks side by side on top of the red block. Directly to the right of these blocks, line up four red blocks end to end as shown.

Suppose you now were to place white blocks on top of these red blocks to complete the wall. How many white blocks would you need?

The answer, of course, is 8. Because you have 2 white blocks for every 1 red block, you would have 8 white blocks for the 4 red ones. Construct the wall to verify that this answer is correct. As you probably know, the relationship between the two walls can be written symbolically as follows:

$$\frac{2 \text{ whites}}{1 \text{ red}} = \frac{8 \text{ whites}}{4 \text{ reds}}$$

or

$$\frac{2}{1} = \frac{8}{4}$$

The = sign means "is the same as."

What you have are two ratios that are equal. This sort of mathematical relationship is known as a *proportion*.

Now place 2 dark green (dg) blocks end to end on the table in front of you. Place 3 purple (p) blocks end to end on top of the dark green blocks. As you see, you have 3 purple blocks for every 2 dark green blocks [a ratio of 3(p)/2(dg)].

Directly to the right of these blocks, place 6 dark green blocks end to end. How many purple blocks would you need to place on top of these dark green blocks to complete the wall?

If you said 9, you were correct. Symbolically, the problem looks like this:

$$\frac{3(p)}{2(dg)} = \frac{?(p)}{6(dg)}$$

One way to solve the problem is simply to note that 2(dg) x 3(dg) = 6(dg), and that 3(p) x 3(p) = 9(p). Or, you could divide 2(dg) into the 6(p) to give 3. Then multiply this 3 by the 3(p) to give 9(p). Construct the wall to verify that this answer is correct.

Start again with 3 purple blocks stacked on top of 2 dark green blocks. Directly to the right of this wall place 9 dark green blocks end to end. How many purple blocks would you need to place on top of these 9 dark green blocks to complete the wall?

This time, the wall cannot be built with a perfect matching of the blocks. You would need 13 1/2 purple blocks to match the 9 dark green ones.

Using the same symbols, this problem would look like this:

$$\frac{3(p)}{2(dg)} = \frac{?(p)}{9(dg)}$$

One way to solve this problem is to divide the 2 dark green blocks into the 9 dark green blocks, giving an answer of 4 1/2 times. Then multiply this 4 1/2 by the 3 purple blocks, which results in 13 1/2 purple blocks.

$$9(dg) \div 2(dg) = 4\ 1/2;\ 4\ 1/2 \times 3(p) = 13\tfrac{1}{2}(p)$$

Construct the wall to verify that this answer is correct.

If you are somewhat unsure of how to solve such problems as these, make up a few problems for yourself with these and other colors of blocks. Try to solve the problems with just a pencil and paper and then verify your solutions by actually building the walls.

Essay 3: The Coupled Gears

For a different experience with proportional relationships, you might like to try to solve some problems with a set of coupled gears. Start by assembling the gears as shown.

Experiment 1

Rotate the smaller gear three times by turning the shaft. How many times did the larger gear turn? Use the table below to record your data.

Experiment Number	1	2	3	4	5	6
Number of Turns of Smaller Gear	3	6		78		2
Number of Turns of Larger Gear			6		5	

Experiment 2

Suppose the smaller gear is rotated six times. How many times do you predict the larger gear will rotate?

Check your prediction by performing the experiment. Record the result in the table.

Experiment 3

How many times do you predict the smaller gear will rotate when the larger gear is rotated six times?

Check your prediction by performing the experiment. Record the result in your table.

As you can see, a direct correlation exists between the number of rotations of the two gears. The more one gear rotates, the more the other gear rotates. In fact, the smaller gear rotates three times for every two times the larger gear rotates. Since the numerical relationship that exists consists of equivalent ratios,

$$\frac{3(s)}{2(1)} = \frac{6(s)}{4(1)} = \frac{9(s)}{6(1)},$$

the relationship is said to be *directly proportional.*

Experiment 4

Suppose the smaller gear is rotated 78 times. How many times do you predict the larger gear will rotate?

Actually performing this experiment will be difficult if you are working by yourself. It would be tedious, and you might lose count. Do you actually need to do the experiment to be sure of your answer?

The problem can be set up like this:

$$\frac{3(s)}{2(1)} = \frac{78(s)}{?(1)}$$

and solved like this:

$$\frac{3(s) \times 26}{2(1) \times 26} = \frac{78(s)}{52(1)}.$$

The 26 was obtained by dividing 78 by 3.

Experiment 5

How many times do you predict the smaller gear will rotate when the larger gear is rotated five times?

Again the problem can be set up like this:

$$\frac{3(s)}{2(1)} = \frac{?(s)}{5(1)}$$

Check your prediction by performing the experiment. Record the result.

Experiment 6

Suppose the smaller gear is rotated only two times. How many times do you predict the larger gear will rotate?

Show how you set up the problem and arrived at your prediction.

Check your prediction by performing the experiment. Record your result.

Essay 4: Balancing Weights

Thus far the correlations you have investigated have all been *direct;* that is, as the value of one variable increased, so did the value of the second variable. But some correlations are *inverse*—as one variable increases, the other variable decreases. See how the variables of distance and weight are correlated in a balance beam.

Start by hanging one of the plastic weights (10 g) at the fourth peg on the right hand side of the beam. Where would you hang another weight to make the beam balance?

If your first guess does not make the beam balance, keep trying until you are successful.

Now try this series of experiments. In each experiment, test your prediction by hanging the second set of weights on the beam.

1. Hang two weights at number 4 on the right. Predict where you would hang one weight on the left to make the beam balance.

2. Hang one weight at number 6 on the left. Predict where you would hang two weights on the right to make the beam balance.

3. Hang one weight at number 3 on the right. Predict where you would hang three weights on the left.

4. Hang two weights at number 3 on the left. Predict where you would hang three weights on the right.

5. Hang five weights at number 6 on the right. Predict where you would hang three weights on the left.

As you have no doubt noticed, the relationship between weight and distance is not a direct one. In fact, the lighter the weight, the farther it has to go out toward the end to make the beam balance. An inverse correlation exists. How can such problems as these be solved?

Look at the first problem. You were given half of the original weight. To balance the beam, you had to hang it out twice as far as the original weight was hung. The original weight was hung at the fourth peg, so the lighter weight had to be hung at the eighth peg (2 x 4 = 8). In short, it's one-half the weight, so it goes twice as far out. Symbolically, the problem looks like this:

1. $$\frac{2\text{(weight on right)}}{1\text{(weight on left)}} \diagdown\!\!\!\!\!\diagup \frac{8\text{(distance on left)}}{4\text{(distance on right)}}$$

The crossed arrows indicate the inverse relationship. The weights and distances are said to be related in an inverse, or reciprocal, fashion.

By the same procedure, the remaining problems can be set up as follows:

2. $$\frac{1\text{(weight on left)}}{2\text{(weight on right)}} \diagdown\!\!\!\!\!\diagup \frac{?\text{(distance on right)}}{6\text{(distance on left)}}$$

It is twice (2/1) the weight, so it must be hung at one-half (1/2) the distance; 1/2 of 6 = 1/2 x 6/1 = 6/2 = 3.

3.

$$\frac{1(\text{weight on right})}{3(\text{weight on left})} \quad \times \quad \frac{?(\text{distance on left})}{3(\text{distance on right})}$$

It is three times (3/1) the weight, so it must be hung one-third (1/3) the distance; 1/3 of 3 = 1/3 x 3/1 = 3/3 = 1.

4.

$$\frac{2(\text{weight on left})}{3(\text{weight on right})} \quad \times \quad \frac{?(\text{distance on right})}{3(\text{distance on left})}$$

It is three halves (3/2) the weight, so it must be hung two-thirds (2/3) the distance; 2/3 of 3 = 2/3 x 3/1 = 6/3 = 2.

5.

$$\frac{5(\text{weight on right})}{3(\text{weight on left})} \quad \times \quad \frac{?(\text{distance on left})}{6(\text{distance on right})}$$

It is three-fifths (3/5) the weight, so it must be hung five-thirds (5/3) the distance; 5/3 of 6 = 5/3 x 6/1 = 30/3 = 10.

Here are a few additional problems you might like to try to be sure that you understand the relationships.

6. Hang three weights at number 10 on the right. Where would you hang five weights on the left? ____

7. Hang four weights at number 2 on the left. Where would you hang two weights on the right? ____

8. Hang three weights at number 8 on the right. Where would you hang four weights on the left? ____

9. Hang seven weights at number 5 on the left. Where would you hang four weights on the right? ____

10. Hang five weights at number 7 on the right. Where would you hang seven weights on the left? ____

Study Questions

1. The graph shows that as the dose of X-radiation increases, so does the number of mutations in organisms. The relationship is directly proportional. For every 5 relative units increase in dosage of X-radiation, 4 mutations occur.

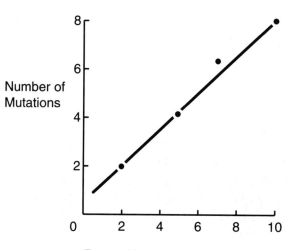

a. Suppose the organisms were given a dose of 30 relative units of X-radiation. How many mutations would you predict would occur? _____

Hint: $\dfrac{5\text{ru}}{4\text{m}} = \dfrac{30\text{ru}}{?\text{m}}$

b. How many mutations would occur if the dosage were 8 relative units? _____

2. The speed at which nerve impulses move is directly proportional to the diameter of the nerve fiber. The speed is about 16 km per hour for every 2 microns diameter of fiber.

At about how many kilometers per hour will a nerve impulse travel down a nerve 16 microns in diameter? Down a nerve 9 microns in diameter?

3. A triple recipe for cookies calls for 4 cups of milk and 5 eggs. How many cups of milk and eggs should you use for a double recipe?

4. John is 6 years old; his sister Linda is 8 years old. When John is twice as old as he is now, how old will Linda be?

5. The time it takes for two types of molecules to move across a cell membrane is directly proportional to the square roots of their molecular weights. In an experiment comparing movement of molecule A (molecular weight 720) and molecule B (molecular weight 370), molecule A is detected inside the cell approximately 6 seconds after its release into the extracellular fluid. How long after release would you expect molecule B to be detected in the cell?

6. A department store is having a sale on socks—6 pairs of socks for $4.00. Another department store also is having a sale on the same kind of socks—8 pairs for $5.25. Suppose both department stores were an equal distance from your house. At which store would you get a better deal? Why?

7. A gardener must mix oil and gasoline for a lawn mower. One-half pint of oil must be added for each one gallon of gas. What are the two variables in this situation? What is the relationship between these variables (constant difference, proportion)?

_____

_____

8. Sam and Mai-Ling are parachuting toward the earth at the same speed. Sam jumped out of the plane after Mai-Ling, so he is above her. What are the two variables in this situation? What is the relationship between them (constant difference, proportion)?

_____

_____

9. A student is measuring the heights of various objects and the lengths of their shadows at a certain time of day. What are the two variables? What is the relationship between them?

_____

_____

_____

10. A distance of 16 km in the metric system is equivalent to a distance of 10 mi in the English system. How many kilometers are equal to 35 miles?

11. In spring, the depth of water in the shallow part of a pond is 60 cm, while at the deep end it is 240 cm. During the summer, water evaporates and the depth of water in the shallow end is only 15 cm. How deep is the deep end in the summer?

12. The unit of currency in Switzerland is the Swiss franc. Fourteen Swiss francs are worth four U.S. dollars. How many francs would you receive for six dollars?

Brain Benders

1. Professor Clearwater, an ecologist, wanted to find out how many frogs live in a pond near the field station. Since she could not catch all the frogs, she caught as many as she could, put a white band around each frog's left hind leg, and then put them back into the pond. A week later she returned to the pond and again caught as many frogs as she could. Here are the professor's data.

> First trip to the pond: 55 frogs caught and banded
> Second trip to the pond: 72 frogs caught, 12 of which were already banded

The professor assumed that the banded frogs had mixed thoroughly with the unbanded frogs, and from her data she was able to approximate the number of frogs that live in the pond. If you can compute this number, do so.

2. Assuming that the earth is round and that the sun is so far away that its rays strike the earth parallel to each other, and given the following information, Eratosthenes of Alexandria (273–192 B.C.) accurately computed the distance around the earth.

a. The distance from the city of Alexandria to the city of Syene is 800 km.

b. Alexandria is directly north of Syene.

c. At noon on June 21, a post casts no shadow at Syene. (The post is pointing toward the center of the earth.)

d. At noon on June 21, a similar post casts a shadow at Alexandria. (This post also points directly toward the center of the earth.)

e. The angle at which lines drawn from the posts to the center of the earth intersect is 7 degrees.

See whether you can do what Eratosthenes did and figure out the distance around the earth.

Investigation 21 *What Is the Structure and Function of Flowers?*

Introduction

Everyone has seen many types of flowers of various sizes and colors from a variety of different plants. But have you ever looked closely at the parts of flowers? Do all flowers have the same parts? What do the various flower parts look like? What are they for? In this investigation, you will examine closely a variety of flowers to discover similarities and differences in their parts and to diagram the parts of a "typical" flower. You also may be able to conduct an experiment to test alternative hypotheses about what the parts do.

Objectives

1. To discover the parts of a flower.
2. To draw a diagram of the parts of a typical flower.
3. To create and discuss alternative hypotheses about functions of flower parts.
4. To design and conduct experiments to test alternative hypotheses. (optional)

Materials

variety of flowers	dissecting microscope	plant cuttings*
metric rulers	compound microscope*	potting containers*
hand lens	droppers*	potting soil*
cutting surface	fertilizer*	seeds or small plants*
dissecting probes	forceps*	slides and coverslips*
razor blades	grow lights*	

*optional materials

Procedure

Part 1

1. Observe and examine a flowering plant. **Caution: Always wash hands after handling plants.**
   a. Describe how the flowers are attached to the plant.

   _____

   b. In what ways does the flower appear to be similar to and different from the rest of the plant?

   _____

   c. Do all the flowers on the plant look the same? Describe any differences.

   _____

2. Obtain and carefully examine flowers from various plants. Carefully dissect the flowers and note similarities and differences in structure. **Caution: Be careful while using sharp instruments.** Sketch your observations.

3. In light of the information you have observed, draw what you think represents a typical flower. Make the flower structures large and clear.

4. Through class discussion, label by name all parts of your typical flower.

5. List proposed functions for the observed flower parts. What evidence, if any, exists to support these hypothesized functions?

_____

_____

_____

_____

_____

_____

_____

_____

_____

Part 2

1. Design an experiment to test your hypothesized functions. Limit your investigation(s) to one or two structures. **Caution: If you use a growlight, make sure electric cords are not in a place where someone can trip over them.**

2. Discuss your experimental design with your teacher.

3. Record on a data sheet your alternative hypotheses, experimental procedure, expected results, and location and common name of your experimental plant.

4. Conduct your experiment. Keep records of your procedure, observations, data, and conclusions.

5. The write-up of your experiment should include the following:
    I.   Question(s)
    II. Alternative Hypotheses
   III. Procedure
   IV. Expected Results
    V. Actual Results (Data)
   VI. Discussion
  VII. Conclusion
In your conclusion, you should be able to answer the question, _Why do plants produce flowers?_
 VIII. Did you conduct a controlled experiment? Explain.

6. Be ready to present your experiment and results to the entire class.

Study Questions

1. Are there such things as male flowers? Female flowers? Flowers that are both male and female? If so, how are they similar? Different? Provide examples of each type if possible.

_____

_____

_____

_____

_____

_____

2. Select one typical flower part that you have not experimented with previously. Describe its structure and propose two possible functions. What experimental evidence exists that the structures actually function in one or more of the ways you state? If you know of no evidence, describe an experiment to test the hypothesized functions.

_____

_____

_____

_____

_____

_____

_____

_____

_____

_____

Investigation 22 *What Is the Function of Fruits?*

Introduction

Fruits are consumed as food by humans and by many other organisms. But do plants produce fruits simply as a food source for other organisms? This seems unlikely. Of what use are fruits to the plants that produce them? You will make some observations that may allow you to answer this question.

Objectives

1. To observe several types of fruits and note their external and internal features.
2. To record differences and similarities among the fruits and speculate on how plants could be classified on the basis of fruit structure.
3. To look for similarities between flower parts and fruit structure.
4. To create alternative hypotheses about fruit function.
5. To design experiments to test alternative hypotheses about fruit function.

Materials

variety of fruits   dissecting needle
metric rulers   forceps
cutting surface   hand lens or dissecting microscope
scalpel or razor blade   bell pepper

Procedure

1. Examine the exterior of the fruit samples provided. List ways in which the fruits vary (e.g., size, shape, color) and list some values of those variables. **Caution: Never taste any fruit samples unless instructed to do so by your teacher.**

   _____

   _____

   _____

   _____

   _____

2. Select some fruits and carefully cut them open. **Caution: Be careful when using sharp instruments.** Consider the following questions:

   a. Do all fruits have similar structures?

   _____

   _____

b. If not, what differences and similarities exist?

c. Does any relationship exist between the presence or absence of internal structures and the external structures observed at step 1?

d. Do all fruits produce approximately the same number of seeds?

e. If not, which fruits or types of fruits produce the most? The least?

f. Do you see any evidence that fruits are similar to flowers? If so, what?

3. Record your observations on a data sheet. You may wish to make sketches of specific fruit structures.

4. What would be a possible method(s) to classify the various fruits into groups?

5. Attempt to sketch a fruit showing "typical" structures.

6. Create alternative hypotheses about what these typical structures are for and think of ways in which your hypotheses could be tested. Be prepared to discuss your ideas in a class discussion.

Study Questions and Additional Activities

Obtain a bell pepper. Open it and count the seeds. Bell pepper plants produce approximately ten fruits each. Use this information to answer the following questions.

1. Approximately how many flowers are formed on a green bell pepper plant?

2. How many seeds were in your bell pepper fruit?

3. How many new pepper plants could be produced from a single fruit?

_____

4. How many new pepper plants could be produced from a single plant?

_____

5. Devise a method to calculate the number of bell peppers that could be produced per acre (43,560 sq. ft. or 1/64 sq. mile) during a growing season.

6. Devise a method to calculate the number of new plants that could be produced if all of the seeds from the above harvest were planted.

7. What variables might affect the biotic potential of the bell pepper plants? Explain.

_____

_____

_____

_____

8. Of the fruits you examined, which seem to be from plants with a relatively high biotic potential? From plants with a low biotic potential? Explain.

_____

_____

_____

_____

Investigation 23 *Where Do Seeds Get Energy?*

Introduction

Living organisms require energy to stay alive and grow. Animals, for example, consume plants and/or other animals, which they digest to obtain energy. Where do plants get the energy they need? Many plants grow from seeds. Is the seed an energy source? Or do seeds ingest other plants and/or animals for their energy? Where do seeds obtain energy needed to sprout and grow into mature plants? In this investigation, you will create and test alternative hypotheses to find out.

Objectives

1. To identify and give names to the major parts of seeds.

2. To create and test alternative hypotheses about the role of seed parts.

3. To gain experience in designing and conducting controlled experiments.

Materials

paper towels
variety of soaked seeds (e.g., beans, peas, corn, sunflower)
dissecting kit (scalpel, probes, scissors, forceps)
iodine solution
sprouting containers
graph paper
labels
metric ruler
balance

Procedure

1. Working with one or two partners, obtain at least six different kinds of soaked seeds. **Caution: Do not eat or taste any seeds.**

2. Using materials found in the dissecting kit, split open each seed to display its internal parts. **Caution: Always use extreme care when handling dissecting equipment.**

3. After observing internal parts, stain each specimen with iodine. Use only a very small amount of iodine; observations will be obscured by any excess. **Caution: Iodine is poisonous. Be careful not to inhale fumes. Iodine will stain skin or clothing.**

4. On a data sheet, sketch each seed and its observable parts. Be prepared to compare your sketches with those of other students in a class discussion. Do all seeds appear to have the same structures?

5. What do you think the role of each seed part might be during the sprouting process? Record your alternative hypotheses on your data sheet.

_____

_____

_____

_____

_____

6. How would you set up an experiment to test your hypotheses about the role of identified seed parts during sprouting? Compare your ideas with your partner(s). Describe your experimental design on your data sheet. Be prepared to discuss your ideas and your experimental design with the entire class.

7. Conduct your experiment to test your hypotheses. Record your expected results, observations, and measurements. Graph your data to show the daily changes that occur during the experiment. Continue the experiment for at least 5 days.

Study Questions

1. In light of your results, what appears to be the role of each seed part during sprouting?

2. What is (are) the energy source(s) for the sprouting seed?

3. Do you think your results are generalizable to all seed plants? Explain.

4. Do you think the energy source(s) for adult plants is (are) the same as that for sprouting seeds? Explain.

5. List ways in which seeds are used by other organisms (including humans).

6. Did you design and conduct a controlled experiment? What were your dependent and independent variables? Which variables should have been held constant?

7. Obtain a copy of the diagram in the back of this manual. Fill in the boxes and blanks with one causal question, hypothesis, experiment, expected result, actual result, and conclusion from this investigation. (See Reasoning Module 5 for details.)

Investigation 24 *What Variables Affect Plant Growth?*

Introduction

Everyone knows that plants will not grow without water. What does the water actually do for the plant? Some people believe that water is a source of food or energy. Do you agree? Perhaps you believe that light is an energy source for plants. If so, what evidence do you have for this? What other sources of energy might a plant have? Surely a variety of factors, such as type of soil and temperature, in a plant's environment will affect its growth and development. Are these also sources of energy used for growth?

The major purpose of this investigation is to try to determine in just what way such variables as type of soil and amount of light affect the growth and development of plants and, if possible, to determine a growing plant's source of energy.

Objectives

1. To discover environmental factors favorable for the germination and growth of selected seeds.

2. To design and conduct controlled experiments to test alternative hypotheses about a growing plant's source of energy.

3. To discover the energy source for growing plants.

Materials

metric rulers
gravel
sand
garden potting soil
seeds (e.g., beans, corn, grass)
refrigerator
plastic bags

labels
light source/heat source
graph paper
paper towels
sprouting containers
water
twist ties

Procedure

1. What is the source of energy for plant growth? List possible energy sources (alternative hypotheses) for plant growth.

_____

_____

_____

2. Think of ways you could experiment to discover which of the possible energy sources is (are) the actual energy source.

3. Describe your experimental design and be prepared to discuss your ideas in a class discussion. Plan to test the effect of at least three independent variables. **Caution: If you use a heat source, keep all electric cords in a place where students cannot trip over them.**

_____

_____

_____

4. Assuming that your hypotheses are correct, what are the expected results of your experiment? Record.

5. Design a data sheet and use it to record your results over the next few weeks. Graph your data.

6. Be prepared to present your data and graph in a class discussion.

7. Do your data and those of your classmates provide evidence to support or contradict the hypotheses advanced? If so, how?

Study Questions

1. What do you now think the source(s) of energy is (are) for continued plant growth? Give evidence.

2. Do plants initially grow at the same rate in light and dark? If not, what reasons can you propose for the differences?

3. What are the differences in appearance between plants grown in the light and in the dark?

4. List as many possible reasons as you can to explain any differences in appearance that you observed.

5. What were the dependent and independent variables in your experiment?

6. Did you conduct a controlled experiment? Explain.

7. Obtain a copy of the diagram at the back of this lab manual. Fill in the boxes and blanks with one causal question, hypothesis, experiment, expected result, actual result, and conclusion from this investigation. (See Reasoning Module 5 for details.)

Investigation 25 *What in the Air Do Plants Need to Grow?*

Introduction

The British chemist Joseph Priestley (1733 – 1804) discovered that a candle will burn under an inverted jar for a short period of time. Then, for some reason, its flame will go out. He also discovered that a small animal, such as a mouse, soon dies if left in similar enclosed spaces, but generally, a plant will survive under the same conditions. Eventually, he came to the conclusion that plants somehow reverse the effect of a burning candle and breathing animals.

Can you suggest a theory to explain Priestley's observations and conclusions regarding plant growth, burning candles, and breathing animals?

Objectives

1. To propose a theory to explain how the growth of plants and animals affects the composition of air.

2. To design and conduct controlled experiments to test the theory.

Materials

bromthymol blue (BTB)   day-old tap water
wood splints   beakers
matches   light source
test tubes   tape
drinking straws   ammonia
pond snails   vinegar
elodea

Procedure

1. Propose a theory to explain Priestley's observations and conclusions. Compare your ideas with those of other students in class discussion.

    _____

    _____

    _____

2. Design an experiment that uses the materials provided to test the theory discussed in class. To do this lab, you will need some background on gas indicators (BTB, glowing wood splint). Your teacher will demonstrate how a change of BTB solution from blue to yellow indicates the addition of CO_2 (and vice versa) and how the flaming of a glowing wood splint indicates the presence of O_2. **Caution: Use care when handling flame or heat sources. Do not inhale fumes from any chemicals. Ammonia is poisonous. Wear your laboratory apron and safety goggles.**

    _____

    _____

    _____

3. Be prepared to discuss your experimental design in a class discussion.

4. List your expected results in the data table below, based on your hypotheses and experimental design.

5. Following completion of your experiment, record your actual results in the Data Table.

Data Table

Test Tube Number	Contents	Original Color	Expected Color Change	Actual Color Change	Expected Splint Reaction	Actual Splint Reaction
1						
2						
3						
4						
5						
6						
7						
8						

Study Questions

1. Did your expected results match what actually happened? If they did match, what statement can be made about your hypothesis?

_____

_____

2. If your expected results did not match your actual results, what changes can be made so that they do match?

_____

_____

3. Did you conduct a controlled experiment? Explain.

_____

_____

4. If your experiment was not a controlled experiment, how could that affect your answers to questions 1 and 2?

_____

_____

5. Years ago, plants were taken out of hospital rooms at night because people believed plants poisoned the air at night. Do you think this practice was based on a good reason? Explain.

_____

_____

6. Obtain a copy of the following diagram. Fill in the boxes and blanks with one causal question, hypothesis, experiment, predicted result, actual result, and conclusion from this investigation. (See Reasoning Module 5 for details.)

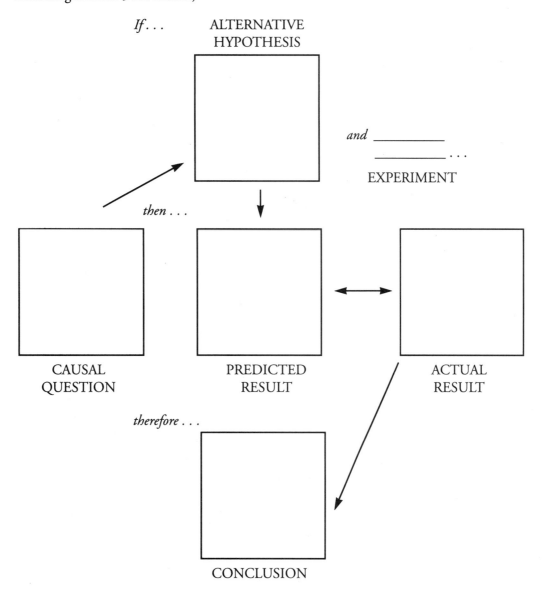

Investigation 26 *How Do Different Parts of Leaves Function?*

Introduction

Plant leaves, and in some cases stems, are able to use the energy of light to make food by combining carbon dioxide and water during a process called *photosynthesis*. In this investigation, you will study the leaf, identify leaf tissues, and then test alternative hypotheses about how these tissues function during photosynthesis.

Objectives

1. To discover leaf structures.

2. To draw a model of a typical leaf.

3. To create and test alternative hypotheses about the functions of leaf tissues.

Part I—The "Model" Leaf

Materials

leaves from 3 different plants
new single-edged razor blade
compound microscope
slides and coverslips
cutting surface

butcher paper
markers or crayons
kalanchoe leaf
dropper

Procedure

1. Obtain a leaf from each of three different plants. What structures can you observe? What similarities and differences exist? **Caution: Never taste or eat any plant or plant parts.** Draw a top and a bottom view of each leaf.

2. Place the leaves on a nonslip surface. With a new single-edged razor blade, carefully cut a paper-thin cross section of each leaf for viewing under the microscope. **Caution: Be extremely careful when handling the razor blade.**

3. Make a wet mount of each section. The cut surface must be facing up.

4. If your cross sections are uniformly thin and are positioned correctly on the slide, observe each cross section first under low and then high power. What microscopic structures can you observe?

5. Make drawings of any structures found.

6. Draw the structures that seem to be common to all of your leaves. Make the drawing bold and clear. Be ready to present your "model" leaf to the class for group discussion.

7. Construct a class model for a leaf. Draw this class model on your data sheet.

8. Participate in a class discussion of different leaf tissues and their possible functions during the process of photosynthesis i.e., *How might each tissue function?* List leaf tissues and hypothesized functions.

_____

_____

_____

Part II —What Is the Function of the Outer Covering?

Materials

kalanchoe leaf
forceps
scalpel or razor blade
petroleum jelly
plastic bag

balance
paraffin wax
light source
covered cabinet or drawer
petri dishes

Procedure

1. What possible functions can you think of for the outer covering of the leaf? List them.

2. Design an experiment that uses the materials provided to test two or more of your alternative hypotheses about the function of this outer covering. **Caution: Be careful when handling the scalpel or razor blade.**

3. Discuss your experimental design with your teacher.

4. Conduct your experiment. Keep a record of your procedure, observations, data, and conclusions.

5. A write-up of your experiment should include the following:
    I. Question
    II. Alternative Hypotheses
    III. Procedure
    IV. Expected Results
    V. Actual Results (Data)
    VI. Discussion of Results
    VII. Conclusion

    In your conclusion, you should answer the question, *What is the function of the outer covering of the leaf?*

    VIII. Did you conduct a controlled experiment? Explain.

7. Be ready to present your experiment and results to the class.

Part III—What Is the Function of the Inside Layers?

Materials

kalanchoe leaf
forceps
distilled water
covered cabinet or drawer
light source
plastic bag
hot plate
petri dishes

iodine solution
dropper
500-mL beaker
large test tube
Benedict's solution
compound microscope
slides and coverslips
safety goggles

Procedure

1. Propose a hypothesis or alternative hypotheses about a possible function or functions of the inner layers of the leaf.

_____

_____

2. [icon] Design an experiment that uses the materials provided to test one or more of your hypotheses about the function of these middle layers. **Caution: Wear your safety goggles. Use care when handling heat sources and chemicals.**

3. [icon] Discuss your experimental design with your teacher.

4. [icon] Conduct your experiment and include the following in a write up.
   I. Question
   II. Alternative Hypotheses
   III. Procedure
   IV. Expected Results
   V. Actual Results (Data)
   VI. Discussion of Results
   VII. Conclusion

   In your conclusion, you should answer the question, *What is the function of the inner layers of the leaf?*

   VIII. Did you conduct a controlled experiment? Explain.

5. Be ready to present your experiment and results to the entire class.

Part IV—What Is the Function of the Leaf Veins?

Materials

beakers, assorted sizes	razor blade
water	potted houseplants
dissecting microscope	fresh celery stalks
food coloring	metric ruler
plastic bags	balance
petroleum jelly	petri dishes

Procedure

1. [icon] Propose a hypothesis or alternative hypotheses about the function of the veins inside the leaf of the plant. **Caution: Never taste or eat plant parts. Some plants are poisonous.**

_____

_____

2. [icon] Design an experiment that uses the materials provided to test one or more of your hypotheses about the function of these leaf veins. **Caution: Be careful when handling the razor blade and glassware. Immediately inform your teacher of any cuts.**

3. Discuss your experiment with your teacher.

4. Conduct your experiment. Keep a record of your procedure, observations, data, and conclusions.

5. A write-up of your experiment should include the following:
   I.   Question
   II. Alternative Hypotheses
   III. Procedure
   IV. Expected Results
   V.   Actual Results (Data)
   VI. Discussion of Results
   VII. Conclusion

   In your conclusion, you should answer the question, *What is the function of the leaf veins?*

   VIII. Did you conduct a controlled experiment? Explain.

6. Be ready to present your experiment and results to the class.

Study Questions

1. Why might succulent leaves be a good adaptation in arid climates?

2. Desert plants tend to have small leaves. Why might this be an advantage to the plant?

3. The Palo Verde is a tree that has green bark. It drops its small leaves during the dry seasons of the year. How might it be able to carry on photosynthesis during these dry periods?

4. Cactus leaves have been modified into spines and are no longer the photosynthetic organs of the plant. How might cacti be able to carry on food production?

5. Explain why the loss of leaves as the photosynthetic structures could be a good adaptation for arid environments.

6. What reason(s) can you propose for the observation that plant roots tend not to be green?

_____

_____

_____

7. When put into a dark area, green plants begin to fade and drop leaves. Propose an explanation.

_____

_____

_____

8. The typical photosynthesis equation states that photosynthesis produces sugar, yet in this investigation you may have tested for starch. Why test for starch, as well as for sugar? Should the equation be rewritten? Explain.

_____

_____

_____

9. For the experiment that you conducted, list the independent variable tested, at least three variables that were controlled, and the dependent variable measured.

_____

_____

_____

Investigation 27 *What Colors of Light Are Used During Photosynthesis?*

Introduction

Experiments that compare plants grown in the light with those grown in the dark provide evidence to support the hypothesis that light is an energy source for photosynthesis and plant growth. But light comes in many colors. Are all colors of light equally effective at driving photosynthesis? If not, which colors are best? In this investigation, you will have an opportunity to experiment to find out.

Objectives

1. To investigate the nature of light.

2. To determine which colors of light are used by plants during photosynthesis.

3. To investigate the nature of plant pigments.

Materials

Part I Spectroscope or Diffraction Grating

plant pigment extract
light source
colored acetate
cardboard box
elodea or hornwort
test tube
beaker
0.25% sodium bicarbonate solution
colored construction paper (e.g., white, red, blue, green, black)
safety goggles
graph paper

Part II Mortar and Pestle

spinach leaves
fine sand
5 mL acetone
cheesecloth
small beaker
jumbo test tube or small bottle with lid or stopper
solvent solution (92% petroleum ether, 8% acetone)
chromatography paper
dropper
safety goggles

 Caution: Acetone and the solvent solution are FLAMMABLE and TOXIC. Do not breathe the fumes. Keep the room well ventilated. Work under a hood if one is available.

Procedure Part I

1. Obtain a spectroscope or diffraction grating, a bottle of plant pigment extracted from spinach leaves, and a light source. Use these materials or any others you might need and are able to obtain to try to answer these questions: *Is light one "thing," or made up of a combination of "things"? What colors of light seem to be absorbed by the plant pigment? What does this suggest about the colors of light used during photosynthesis?* The pigment is green. Does this suggest that green light is used during photosynthesis? Explain your answer to this last question. Record your answers, observations, procedures, and arguments.

_____

_____

_____

_____

_____

2. Cover your light source with a piece of colored acetate to produce light of a particular color. Shine the light on sheets of construction paper of various colors and note the color of the lighted paper.

3. Select another color of acetate and repeat step 2. Keep track of the color of the light source, the original color of the paper, and the resulting color of the lighted spot on the paper.

4. Try to explain the results. Why does white paper appear white, red, green, and so on, depending on the color of the light used? Why does the black paper always appear black?.

_____

_____

_____

5. Your explanation should be consistent with that in question 1. Is it? If not, you will need to create another explanation.

6. Do plants use all "components" of light equally well? To find out, investigate the effect of the various color filters on the photosynthetic rate of elodea or hornwort. A fresh, growing tip 2 to 3 cm long will produce small oxygen bubbles from the cut stem as an indicator of photosynthetic activity.

7. Observe bubble production under various colors of light for a few minutes. What variables besides color of light might affect the rate of bubble production? Keeping these variables in mind, design and conduct a controlled experiment that will yield quantitative data and will allow you to answer this question: *What colors of light are used during photosynthesis?* Record your design. The 0.25% sodium bicarbonate solution maybe used as a source of CO_2 for the plant.

_____

_____

_____

_____

_____

8. If you have an idea how this experiment will turn out (what do you expect will happen?), record your ideas, along with your reasons for your expected results.

_____

_____

_____

_____

9. Conduct your experiment. Record your data. Graph results if possible and be prepared to present your data during a class discussion.

Procedure Part II

1. Do plants use a single type of pigment molecule to absorb light energy, or do they use a variety of pigments? Paper chromatography is a technique that can be used to separate different types of molecules from one another. In this activity, you will use paper chromatography to discover whether mashed spinach leaves contain more than one type of pigment molecule. If so, it may be that plants (spinach at least) use a variety of pigments for photosynthesis.

2. ☠ Obtain three spinach leaves, fine grinding sand, and approximately 5 mL of acetone. **Caution: To avoid harmful toxic fumes, work in a well-ventilated area or under a hood.** Place these materials into a mortar and grind with the pestle until a dark, pulpy mass is formed. (You may need to ask your teacher for more acetone, as it evaporates quickly.)

3. Filter the mixture through cheesecloth, collecting the filtrate in a small beaker.

4. Obtain two strips of chromatography paper. (Why shouldn't you use a single strip?) Handle the paper only by the edges so that you will not get oils from your fingers on the face of the paper.

5. Place a small drop of the plant pigment filtrate in the center of each strip, approximately 3 cm from the bottom. Let dry completely.

6. ☠ Obtain a large test tube or bottle and pour 2 cm of the solvent solution into the bottom. **Caution: The solvent is flammable and toxic. Do not breathe the fumes.**

7. Insert the chromatography strips into the test tube, submerging the bottoms so that the spots of pigment are 7 to 8 mm above the liquid. Clamp the top of the strips under the stopper or lid.

8. Seal the vessel and keep it perfectly vertical.

9. Allow the solution to soak to very near the top of the strips. As the solution is rising, clean up your work area according to your teacher's direction.

10. Remove the strips and record what you see. Compare your strips with those of other students. How are they similar? Different?

_____

_____

_____

11. What conclusions can be drawn?

_____

_____

_____

Study Questions

1. Do plants contain more than one pigment? What is your evidence? Does this indicate that plants use more than one pigment to capture light? If not, what further evidence is needed?

2. Why do you think most plant leaves are green like those you tested?

3. How do the observations made in this investigation help to explain the color changes seen in the leaves of broad leaf plants in the fall?

4. Carrots contain a pigment that gives them their distinct orange color. Did you see any evidence that the plants you tested contained any of this pigment? Explain.

5. Did you conduct controlled experiments in this investigation? Explain. What were the independent and dependent variables? What variables were held constant?

Investigation 28 *Is Chlorophyll Necessary for Photosynthesis?*

Introduction

In Investigation 24 you discovered that only plants grown in light turn green and continue to grow. Such observations as this support the hypothesis that light is a necessary energy source for the process of photosynthesis and plant growth. Could the green color in plants also be necessary for photosynthesis? The green color is presumably due to pigment molecules called *chlorophyll* that absorb certain wavelengths of light and reflect others. If chlorophyll is really necessary, then any plant without it should not be able to conduct photosynthesis.

In this investigation, you will grow corn seeds, some of which will produce albino plants with no chlorophyll. You then will compare the ability of the albino plants and normal green plants to produce starch. Starch is a complex combination of sugar molecules and thus a major product of photosynthesis. If chlorophyll is really necessary for photosynthesis and starch production, then normal green plants with chlorophyll should be able to produce starch but albino plants with no chlorophyll should not.

Objectives

1. To test the hypothesis that chlorophyll is necessary for photosynthesis in corn plants.

2. To examine plant leaf cells microscopically to look for the location of chlorophyll molecules.

Part I—Growing Corn

Materials

corn seeds (6 albino, 6 normal)   water
scissors   light source
plastic-foam egg carton   labels
potting soil   waterproof marker
drain tray   metric ruler

Procedure

1. State the hypothesis that this investigation is designed to test.

   _____

2. State any alternative hypotheses you may be able to think of regarding the role of chlorophyll.

   _____

   _____

3. To test these hypotheses, you will need some growing albino and normal corn plants. Cut the top half off an egg carton. Discard or recycle this top half. Cut the bottom part of the egg carton into two equal halves. You will be planting corn seeds in one of these halves. Give the other empty half to another group to use. **Caution: Handle scissors with care.**

4. Punch a hole with your pencil through the bottom of each egg slot. This hole ensures proper drainage. Fill the egg slots with potting soil. Plant two corn seeds in each egg slot (a total of 12 seeds). Set your potted seeds into a drain tray, water the seeds, and put the plants where they will get plenty of light. **Caution: Never eat any seeds or plant parts.**

5. State an expected result derived from your major hypothesis and the experimental design.

_____

6. Construct a data table and graph the growth of your plants over a 3-week period. Be prepared to put your graph on the board for class discussion. Consider the following questions:

   a. In what ways do all the corn plants look alike?

   _____

   _____

   b. In what ways do the plants look different?

   _____

   _____

   c. Which of the plants came from the albino seeds? Which came from the normal seeds?

   _____

   _____

   d. How does the growth pattern differ between the albino and normal plants?

   _____

   _____

   e. What may be some causes of these differences; that is, are any of your hypotheses supported or contradicted? Explain.

   _____

   _____

Part II—Looking at Corn and Elodea More Closely

Materials

compound microscope
dissecting microscope
slides and coverslips
scalpel or razor blade

cutting surface
dropper
corn plants (green and albino)
elodea plants

Procedure

1. Use the materials to try to answer these questions: *Where is the chlorophyll located inside the corn plants? Do the albino and normal corn plants show any structural differences? If so, what are they?* Record your observations. *Where is the chlorophyll inside elodea plants? Does this appear to be the same place as in the corn?*
   Caution: Be careful when handling the scalpel or razor blade. Do not eat any plant parts.

2. Do your observations provide further evidence to support or refute your hypotheses about the role of chlorophyll in photosynthesis and plant growth? Be prepared to share your observations and arguments in a class discussion.

_____

_____

_____

_____

_____

_____

Part III—Chloroplasts and Starch

Materials

safety goggle	corn plants (green and albino)
large test tube	2 petri dishes
ethyl alcohol	dropper
500-mL beaker	iodine solution
water	slides and coverslips
hot plate	compound microscope
forceps	

Procedure-Starch Test

Note: Iodine solution is used to indicate the presence of the complex plant sugar called *starch*. In a positive test for starch, a blue or blue-black color is produced. In a negative test, the iodine solution stays the same reddish color.

1. Do albino and normal corn plants contain different amounts of starch? To find out, fill a large test tube two-thirds full with ethyl alcohol. **Caution: Do not inhale any fumes. Wear your safety goggles.**

2. Half-fill a 500-mL beaker with water. Heat the water to boiling. **Caution: Be careful when placing beakers in water bath. Use forceps.**

3. Put the large test tube into the boiling water bath.

4. Use the forceps to put a leaf from a healthy green corn seedling into the hot alcohol. What gradually happens to the leaf?

_____

_____

5. When no color is left in the leaf, use your forceps to remove the leaf from the alcohol. Put the leaf into a petri dish.

6. Use a dropper to cover the leaf with iodine solution. **Caution: Iodine is poisonous and will stain skin and clothing.** What happens to the leaf?

_____

_____

7. Repeat the test with a leaf from the albino corn plant. What did you observe this time? Are the results similar or different for the albino and normal plants? Does this observation support or refute any of your hypotheses about the role of chlorophyll in photosynthesis? Explain.

_____

_____

_____

_____

_____

Study Questions

1. Where in the elodea leaf did you observe the green color? What is the name of this structure?

_____

_____

2. How did the albino and normal corn plants react to the iodine test? What do these reactions indicate about the role of chlorophyll in plant growth?

_____

_____

3. As a class, construct a table to record the color frequency (green : white) of the corn seedlings. What is the ratio of green to white seedlings? Was the environment of all plants the same? Does this indicate why some of the corn plants are green and some are white? Explain.

_____

_____

4. Assume that chlorophyll is necessary for photosynthesis. How can you explain the initial growth of the albino corn seedlings? What was their energy source?

_____

_____

5. Name the independent and dependent variables in your experiments. Did you conduct controlled experiments? Explain.

_____

_____

6. Obtain a copy of the diagram at the back of this lab manual. Fill in the boxes and blanks with one causal question, hypothesis, experiment, expected result, actual result, and conclusion from this investigation. (See Reasoning Module 5 for details.)

Investigation 29 *What Causes Water to Rise in Plants?*

Introduction

If you place a plant such as a stalk of celery (with leaves) into a beaker with colored water, you soon will notice that the colored water somehow moves up through the celery stalk and into the leaves. Such observations as this suggest that the general pattern of water movement in plants is from the roots, through the stem, to the leaves. But what causes the water to move upward? This movement is against the force of gravity, which pulls things down. Do you have any ideas?

Objectives

1. To create and test alternative hypotheses about the cause(s) of water rise in plants.

2. To identify some of the structures through which water travels in plant stems.

Materials

Schilling red food coloring
test-tube rack
slides and coverslips
compound microscope
variety of plant stems (e.g., bean, celery, coleus, corn, impatiens, orange, paloverde, pyrocantha, sunflower)

colored pencils or markers
test tubes
petroleum jelly
new single-edged razor blade

Procedure

1. *What causes water to rise in plants?* List any ideas you and your classmates have about the cause or causes of the upward movement of water through plants.

   _____

   _____

   _____

   _____

2. Design experiments that use the materials provided to test one or more of these alternative hypotheses. In general, you will have to place plants or plant parts into containers partially filled with colored water and wait several minutes to observe the movement or lack of movement of the colored water through the plant. Try to contradict or support each of the hypotheses by comparing expected results with actual results. Discuss your experimental designs with others in your group or in a class discussion before you proceed. Should you include some sort of control? If so, what and why? **Caution: Use care handling test tubes, glassware, and the razor blade. Never eat any plant parts. Some plants are poisonous.**

3. Summarize your work for each experiment in a table. Use the following heads: Hypothesis; Experimental Manipulation including (independent variable tested); Predicted Results; Actual Result; Conclusion.

4. Were you able to tell precisely where in the plant stem the water was moving? If not, you may want to make some cross sections of plant stems that have had colored water passing through them. Perhaps the colored water will have stained the water-conducting portion of the stem and will be visible in cross section under a microscope.

5. To make a cross section, place a stem on a nonslip surface and cut across it with a new single-edged razor blade. A slicing motion usually causes less crushing than a direct downward push of the blade. Make sketches and/or notes concerning your observations.

6. Be prepared to report your hypotheses, expected results, actual results, and tentative conclusions to the class.

Study Questions

The next seven items are based on the following experimental design:

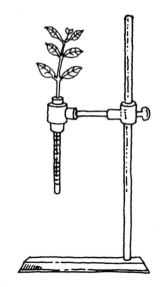

The cut end of a stem is inserted into a glass tube containing water. Movement of water into the plant is shown by a rise of water in the glass tube. This apparatus is called a *transpirometer.*

Five transpirometers were set up and labeled. Use the labels as a key to answer the questions that follow.

KEY:

Transpirometer A: The setup as shown was placed near a sunny, warm window.

Transpirometer B: The setup as shown was placed in a cool, dark room.

Transpirometer C: All the leaves were cut off the plant, and the setup was placed near a warm, sunny window.

Transpirometer D: The setup as shown was placed in a warm, dark room.

Transpirometer E: All leaves were cut off the plant, and the stem was inserted into the glass tube upside down and placed near a warm, sunny window.

1. Name the independent variables in the above setups.

2. Name the dependent variable(s).

3. Why is it improper to compare Setup A with Setup B to test the role of light in transpiration?

4. Which setups can be used to test the idea that light increases the rate of transpiration?

5. Which setups can test the idea that leaves play a role in transpiration?

6. Suppose you found the rate of water movement was greater in Setup C than in Setup B. What reason(s) can you give for this result? Is this a controlled experiment? Why or why not?

_____

_____

7. Name all of the controlled comparisons that can be made with the five setups.

_____

_____

8. Obtain a copy of the diagram at the end of this lab manual. Fill in the boxes and blanks with one causal question, hypothesis, experiment, expected result, actual result, and conclusion from this investigation. (See Reasoning Module 5 for details.)

_____

_____

Investigation 30 *What Could It Be?*

Introduction

Imagine that while walking in a forest you spot a fuzzy, ball-shaped object lying on the ground beneath a large pine tree. After picking it up and carefully looking it over, you conclude that this is something you have never seen before. What could it be?

Objectives

1. To carefully pull apart and observe an unknown object to identify its contents.

2. To arrange selected contents in an orderly fashion to identify their structure.

3. To create and discuss alternative hypotheses about where the unknown object came from.

Materials

unknown object (wrapped in foil)   metric ruler
forceps   glue
dissecting probe   cardboard
plastic sandwich bags   diagram of human skeleton

Procedure

1. *What is the fuzzy, ball-shaped object? Where could it have come from?* After observing the exterior of the unknown object but before you pull it apart, create alternative hypotheses about what you think it might be. Record these ideas.

_____

_____

_____

_____

2. Use forceps and a dissecting probe to carefully pull apart the contents of the unknown object. Observe them carefully. **Caution: Handle the dissecting probe carefully.**

3. Sort the contents into separate groups. What characteristics did you use for your sorting? What might these characteristics tell you about the origin of the contents? Use plastic bags to store the contents. **Caution: Wash your hands thoroughly when you have finished sorting.**

_____

_____

_____

4. Create alternative hypotheses about what the contents are and how they got into the fuzzy ball. Record your hypotheses and compare them with your ideas in step 1.

_____

_____

_____

5. Be prepared to discuss your hypotheses and any evidence you may have to support or refute them in a class discussion.

6. If possible, collect one complete skeleton from the fuzzy ball.

7. Attempt to arrange all bones in their proper position and glue them to a piece of cardboard. **Note:** *To do this, your arrangement will have to be limited to two dimensions.*

8. Use a labeled diagram of a human skeleton to locate the corresponding bones in your specimen. Label the parts of your specimen.

vole

mouse

9. Using the mammal key, attempt to name your specimen and as many other specimens as possible.

Key for Mammal Skulls
1. a. Teeth present. Go to 2
   b. Teeth not present. Bird

2. a. Teeth approximately the same size. Go to 4
   b. Four large front teeth with no color on teeth. Rabbit
   c. With color on teeth. Go to 3

3. a. Skull almost half as thick as it is long. Vole
   b. Skull approximately one third as thick as it is long. Mouse

4. a. Skull is 32 mm in length or longer. Go to 5
   b. Very small, slender skull. Shrew

5. a. Skull length between 32 mm and 40 mm. Rat
   b. Skull over 40 mm long. Mole

Study Questions

1. Construct a food web diagram showing the feeding relationships implied by your observations of the fuzzy ball.

2. How many skulls (or pairs of jaw bones) were in the fuzzy ball?

3. Assuming that one ball is produced per day, estimate the number of prey animals eaten by the predator per year.

4. What might happen to the prey population if the predator population decreased?

5. From the point of view of a farmer, why might predators such as owls be harmful? Beneficial?

_____

_____

_____

6. Why do chemicals designed to kill small insects pose a threat to predators at the top of food webs?

_____

_____

_____

7. Why are food chains usually limited to three or four trophic levels?

_____

_____

_____

Investigation 31 *What Happens to Dead Organisms?*

Introduction

Are the apples you buy in the store dead? Do you suppose the seeds from these apples will grow if you planted them? How would you decide that something is dead? What happens to dead organisms? In this activity, you will observe the breakdown of dead organisms and attempt to discover what factors influence the rate at which breakdown occurs, as well as discovering what actually causes the breakdown.

Objectives

1. To discover factors that affect the rate of breakdown of dead organisms.

2. To create a theory to explain how and why the breakdown occurs.

Materials

plants or plant products (e.g., banana, lettuce, apple, pear, leaf, bread)
dead animals and animal products (e.g., smelt, cricket, mealworm, bologna, hamburger)
sand
soil
antiseptic solution
water
light source
heat source
cold source (refrigerator)
salt
dissecting microscope
alcohol
petri dishes
sugar
vinegar

Procedure

1. Work in lab teams as instructed by your teacher. **Caution: Be sure to wash your hands and work area as instructed.**

2. Use the materials listed or additional materials to set up an experiment or a series of experiments to answer the question, *What independent variables influence the rate of breakdown of dead organisms?* **Caution: Use care when handling glassware and chemicals. Notify your teacher immediately of any spills or breakage. Keep all electrical cords out of the way.**

3. As you set up your experiments, keep a record of what you have done and list any specific questions your experiments attempt to answer. Investigate the influence of at least three independent variables. One of these variables must be amount of water poured on the dead organism. List the three variables you will test. What variables should be held constant?

What is/are the dependent variables in your experiment?

4. During the next few days make regular observations and record them on your data sheet. Remember that you have more than one sense to make observations.

5. Use a microscope and your imagination to help yourself create a theory to answer the question, *What actually causes the breakdown of dead organisms?* Record your postulates of your theory.

Study Questions

1. Why will a box of strawberries spoil but an unsealed jar of strawberry jam will not?

2. Why do canned foods not spoil?

3. Can you think of any environments in which decomposition is slowed or does not occur at all? Explain.

4. Can you think of any environments in which decomposition is greatly enhanced? Explain.

5. Generally, all microbes are thought of as harmful (causing disease or decomposition). Do you know of any instances in which microbes are helpful? If so, what?

6. In light of the class discussion, describe favorable conditions for microbial growth.

_____

_____

7. How do molds reproduce?

_____

_____

8. What happens to the decomposer when it has completely consumed the dead organism?

_____

_____

9. How is the substrate enriched by biological decomposition?

_____

_____

10. What would happen to plants if decomposers were completely eliminated? To animals?

_____

_____

Investigation 32 *What Is the Pattern of Population Growth and Decline?*

Introduction

Populations of all living things are influenced by such environmental variables as temperature, amount of light, amount and type of food, amount of water, and the size and types of other populations of plants and animals. The size of a given population will vary because of changes in these and other environmental variables. For example, when a deer herd has outgrown its available food supply, the individual deer will become undernourished. Some deer may die from starvation.

In this activity, you will investigate the growth and decline of yeast populations under various environmental conditions to try to answer the following questions.

- What is the general pattern of yeast population growth and decline?
- What is the effect of specific such environmental variables as amount of space and amount of food supply on yeast population growth and decline?

Objectives

1. To develop a sampling method to estimate the size of a population of yeast cells in a liquid culture.

2. To discover patterns of population growth and decline.

3. To discover the influence of a number of variables (e.g., amount of space, amount of food, initial population size, etc.) on patterns of population growth and decline.

4. To propose alternative hypotheses to explain observed patterns of population growth and decline.

Materials

yeast cultures in test tubes	caps for test tubes
droppers	10% sugar solution
slides and coverslips	light source
compound microscope	incubator
metric rulers	refrigerator
10-mL graduated cylinders	test tubes, various sizes
50-mL beakers	other sources of environmental variables

Procedure

1. Use the available materials to try to answer these questions: *What is the pattern of yeast population growth and decline? What environmental variables affect yeast population growth and decline?* Choose an environmental variable that you think may affect the growth of the yeast population and design an experiment to test its actual effect. Diagram your experimental design, noting independent and dependent variables and which ones should be held constant. You may wish to check with your teacher before proceeding with your setup. **Caution: Handle glassware carefully. Notify your teacher of any breakage. If using electrical equipment, keep cords out of the way.**

2. After you have set up your experiment, you must determine the initial size of the yeast population in your test tubes. To do this:
   a. Stir the contents of the test tubes gently but so that the yeast are evenly distributed. Place one drop on a slide and cover with a coverslip. Do not apply pressure to the coverslip.
   b. Count the number of yeast cells within the area of the high power field.
   c. Work with your lab team to develop a procedure for accurately estimating the population of

yeast cells in the test tubes. Record your procedure.

3. Record your estimate of the initial (Day 1) yeast population size on your data table (See example table below).

Yeast Population

		Tube Number/Condition			
Day	**Date**	**1/**	**2/**	**3/**	**4/**
1					
2					
3					
4					
5					

4. Store the tubes of yeast culture as instructed by your teacher.

5. On the following days, repeat step 2 just as you did on Day 1 of this experiment. (Remember to stir the test tubes.) Record your estimates in your data table.

6. Graph your data and be prepared to discuss your data and those of your classmates in a class discussion.

7. Consider the following questions:

   a. In comparing your team's data with those of other teams, what similarities and differences did you note? What might be some causes of these differences?

   _____

   _____

   _____

   _____

   b. Using the class data curves, describe the major phases a yeast population seems to go through.

   _____

   _____

   c. Try to explain what causes the changes in a yeast population as it goes from one phase to the next.

   _____

   _____

   d. List environmental factors that appear to be limiting the growth of the yeast population.

   _____

   _____

   _____

   _____

Study Questions

1. What similarities and differences do you think might exist between yeast population growth in this lab and human population growth in the world?

2. What are or will be major factors limiting human population growth in the area in which you live?

3. Below are data on the U.S. population size from 1630 to 1990. Plot these data on a graph.

Population of the United States

Year	Thousands of People	Year	Thousands of People
1630	4	1830	12,901
1650	50	1850	23,261
1670	111	1870	39,905
1690	210	1890	63,056
1710	331	1910	92,407
1730	629	1930	123,188
1750	1,170	1950	151,683
1770	2,148	1970	203,185
1790	3,929	1990	248,710
1810	7,224		

4. Does your graph suggest an arithmetic or a geometric population growth? Explain.

5. By extending the population curve on the U.S. population graph, what do you estimate the population of the United States will be by the year 2050?

6. Do you think food has ever been a limiting factor in the population growth? Explain.

7. Do you think food might become a limiting factor to our population growth in the future? Explain.

8. In light of your prediction in question 5, consider what life will be like in the United States in the year 2050.

    a. Will enough recreational space (parks, forests, clean rivers) exist for you and your family?

_____

_____

    b. List any resources you think might be in short supply and note any you think might become limiting factors to United States population growth by the year 2050.

_____

_____

_____

_____

9. Consider your answers to questions 3 – 8. What suggestions can you make to the leaders of our country?

_____

_____

_____

_____

_____

_____

_____

_____

_____

_____

_____

_____

_____

_____

_____

_____

Investigation 33 *What Causes Population Size to Fluctuate?*

Introduction

The growth of populations of organisms, such as deer, is limited by such factors as predation, limited food and water supplies, and extreme temperatures. Do populations, once they reach their environment's carrying capacity, remain the same size, or do their numbers fluctuate through time? Actually, few populations remain the same size. Most show distinct fluctuations. What causes these fluctuations? In this activity, you will play the roles of deer, food, water, and shelter to simulate the interactions of a deer population with specific biotic and abiotic environmental factors. By analyzing the results of the simulation, you should better understand the reasons for population size fluctuation.

Objectives

1. To discover patterns of animal population size fluctuation in a simulated environment.

2. To propose causes of population size fluctuation.

Materials

area to participate in simulation (outdoors or indoors)
string, rope, or chalk to mark lines
3 notebooks or clipboards with data table (see Table 1)

Procedure

1. Consider a deer population living in a mountain range in the western United States. The size of the deer population will vary from season to season and from year to year. What reasons can you suggest for this? List your alternative hypotheses and those of your classmates.

_____

_____

_____

_____

2. To begin the simulation, your teacher will assign you a number from 1 to 4. Record your number. If your number is 1, you will begin the simulation as a member of the deer population. If your number is 2, 3, or 4, you will begin the simulation as either food, water, or shelter for the deer.

3. All of the "deer" students are to assemble in one location, while all of the "food," "water," and "shelter" students are to assemble in another location.

4. For the purposes of this activity, you are to emphasize and examine the effects of food, water, and shelter on the deer population; therefore, assume that the deer have enough space and other variables. The deer need to find food, water, and shelter in order to survive.
   • When a deer is "looking" for food, it clamps its hands over its stomach.
   • When a deer is "looking" for shelter, it holds its hands together over its head.
   • When a deer is "looking" for water, it puts its hands over its mouth.

5. A deer can choose to look for any one of its needs during each round of activity. A deer cannot, however, change what it is looking for (when it sees what is available) during that round. It can change what it is looking for in the next round, if it survives.

6. Each food, water, or shelter student gets to choose at the beginning of each round which component he or she will be during that round. Students depict which component they are in the same way the deer show what they are looking for; that is, hands on stomach for food, and so on.

7. The game starts with all players lined up on their respective lines (deer on one side; habitat components on the other side) and with their backs to one another. The teacher begins the first round by asking all of the students to make their signs—each deer deciding what it is looking for, and each habitat component deciding what it is.

8. Before each round, a student or the teacher records the number of deer and the total number of food + water + shelter components on the data sheet.

9. • When everyone is ready (backs still turned), the teacher counts "One . . . two . . . three." At the count of three, each deer and each habitat-component turns to face the opposite group, continuing to hold its sign clearly.
   • When a deer sees the habitat component it needs, it runs to that component. Each deer must hold the sign of what it is looking for until getting to the habitat component student with the same sign.
   • Each deer that reaches its necessary habitat component takes the "food," "water," or "shelter" back to the deer side of the line. This is to represent that the deer has successfully met its needs and successfully reproduced as a result.
   • Each "food," "water," and "shelter" person captured then becomes a deer.
   • Any deer failing to find its food, water, or shelter dies and becomes part of the habitat; that is, in the next round, the deer that died are habitat components and so are available as food, water, or shelter to the deer who are still alive.
   • If more than one deer reaches a habitat component, the deer who gets there first survives.
   • Habitat components stay in place on their line until a deer needs them. If no deer needs a particular habitat component during a round, the habitat component just stays where it is from round to round.

10. After the deer have died or have survived and reproduced and the new population has been established for the next round, record their numbers in Table 1 on page 157.

11. At the end of 15 rounds or when instructed by your teacher, gather your materials, return to the classroom, and prepare a line graph of deer population size and total number of habitat components for each of the 15 rounds.

12. What pattern(s) does your graph reveal? What factor(s) seem(s) to be causing these fluctuations?

_____

_____

_____

_____

13. What effect(s) do you think predators might have on population-size fluctuations?

_____

_____

To find out, you may wish to repeat a few rounds of simulation, including predatory mountain lions that can capture and eat deer. First, allow deer to capture food, water, or shelter. Then, allow the mountain lions to capture the deer. Only deer that have previously captured shelter are safe from predation. Predators who fail to capture a deer become food, water, or shelter. Deer who get eaten become predators. Record numbers of mountain lions, deer, and food, water, and shelter components over a number of rounds. Graph the data.

14. Consider the following questions:

- Is it accurate to say that the mountain lion population controls the deer population? Or does the deer population control the mountain lion population? Or do plant populations and abiotic environmental factors control the animal populations? Explain.

_____

_____

_____

_____

- In terms of the simulation, how would you define the phrase "balance of nature"?

_____

- In the simulation, some things changed into other things (e.g., water changed into deer, deer changed into predator). Explain in what ways this is similar to and different from what takes place in nature.

_____

_____

_____

- In what other ways do you think the simulation was similar to and different from nature?

_____

_____

_____

Table 1

Round Number	Food + Water + Shelter	Deer	Mountain Lions
1			
2			
3			
4			
5			
6			
7			
8			
9			
10			
11			
12			
13			
14			
15			

Study Questions and Additional Activities

1. The data in Table 2 show the number of snowshoe hare and lynx pelts (in thousands) captured by trappers across Canada and the northern United States Territories from the year 1875 to 1915. Graph the data on a line graph.

2. Does your graph suggest a relationship between the two populations? If so, what is it?

3. Which animal do you suspect is the predator? Explain.

4. Which animal do you suspect is the prey? Explain.

5. Do you think the predators are controlling the prey, or are the prey controlling the predators? Both? Neither? Explain.

6. Make a prediction about what would happen initially to the predator population if the prey population were to increase suddenly. Explain.

7. Make a prediction about what would happen initially to the prey population if the predators were to be exterminated. What might happen over many generations? Explain.

8. Describe the general relationship that seems to exist between the prey population and the predator population over time.

9. How does *interspecific competition* differ from *intraspecific competition?* Define the terms and give an example.

Table 2
Hudson Bay Company Trapper Data

	Snowshoe Hare (*Lepus americanis*) thousands of pelts	Lynx (*Lynx canadens*) thousands of pelts
1875	102	39
1877	84	48
1879	10	22
1881	8	15
1883	50	42
1885	133	78
1887	116	60
1889	15	28
1891	32	15
1893	72	45
1895	86	50
1897	34	39
1899	2	10
1901	6	12
1903	70	43
1905	52	62
1907	25	3
1909	39	12
1911	50	20
1913	48	42
1915	24	39

Investigation 34 *How Do Organisms Interact With Their Environment?*

Introduction

In this field study, you will investigate the interactions of organisms with each other and with their physical environment. The physical environment includes all the nonliving conditions of the area, such as the amount of sunlight, the humidity, precipitation, temperature, soil type, and the land forms themselves. As these conditions change, will the number and type of organisms change too? What interactions occur among organisms? Are these interactions similar from one type of environment to the next? Or do they also vary? If so, why? If not, why not?

Objectives

1. To discover what organisms live in a specific habitat.
2. To discover special characteristics of organisms that enable them to live successfully in that habitat.
3. To discover differences in plant species and frequency as one moves from a dry stream bed or stream to the surrounding area.
4. To identify environmental factors responsible for these differences.
5. To construct a chart that shows major food chains and food webs of the habitat.

Materials

thermometers   metersticks or metertapes
sling psychrometers   poking stick
shovels   binoculars
string   pencil
nails   transect rope

Procedure

1. After briefly exploring the study area, construct a list of variables of the physical (nonliving) environment that you think might affect organisms in the area. Use any of the equipment available to measure the values of each physical variable. Complete Table 1 at this time.

Table 1

Physical Variable	Value	Predicted Variation Throughout the Day	Predicted Varation Through the Year

2. Identify the different kinds of plants present in the study area and complete Table 2 by drawing the distinguishing features for each. In what way(s) might these characteristics help the plant survive in its habitat?

Table 2

Name of Plant	Distinguishing Features (Sketch)	Predicted Survial Value

3. 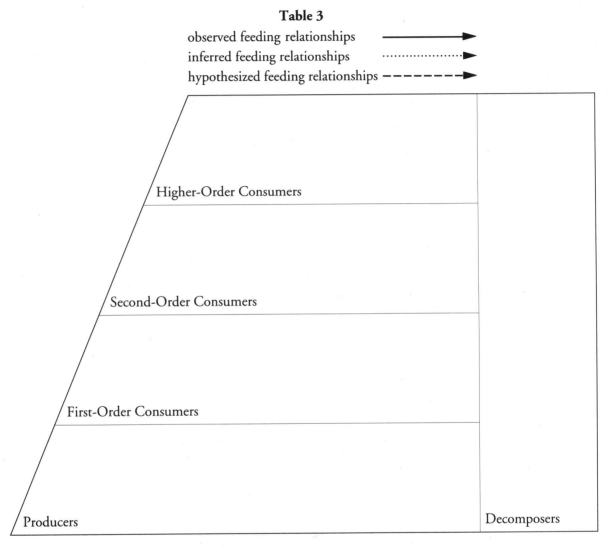 Explore the area for animals and any evidence of animal activity and feeding relationships. Also attempt to find evidence of decomposition. Use your poking stick to turn over objects (e.g., rocks, logs) and to get into small places. **Caution: Be careful not to come into direct contact with poisonous insects or reptiles. Do not extend your hands or feet into places you cannot see.** Find a high spot from which you can use the binoculars to survey the area around you.

4. *What feeding relationships exist among the organisms in the area?* Use Table 3 to record observed, inferred, or hypothesized feeding relationships. Draw arrows to indicate direction of energy transfer.

Table 3

observed feeding relationships   →

inferred feeding relationships   ·············▶

hypothesized feeding relationships –––––▶

Higher-Order Consumers

Second-Order Consumers

First-Order Consumers

Producers   Decomposers

5. Does plant diversity vary in the study area? At a designated time, your teacher will call the groups together to gather data on plant diversity as you move away from the stream or dry stream bed to the surrounding area. Ecologists have determined that a representative sample of plant diversity and distribution can be made by using the line-transect method of alternating quadrats. The kinds and numbers of plants will be determined within each 5 m^2 square, running along the transect line or rope.

Look at the illustration. Note that each quadrat has been numbered starting with *1* in the middle of the stream.

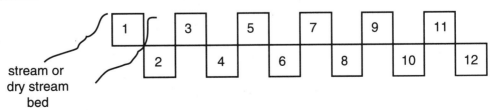

6. Your teacher has laid out the transect rope for you. Verbal instructions will be given to you on marking off the 5 m² square quadrats with string and nails.

7. Your teacher will assign your team a plant species to count in each of the quadrats 1 to 12 along the transect rope. Complete Table 4 with your results. Include data on the approximate area covered by your species.

8. Design graphs to illustrate your data from your Table 4.

9. After graphing your data from Table 4, do you see any patterns of plant distribution? If so, describe.

_____

_____

10. What environmental variables might be responsible for these differences?

_____

_____

11. What characteristics and/or behaviors do the animals you found have that enable them to survive in their particular habitat?

_____

_____

_____

_____

_____

12. What evidence did you find for interactions among organisms other than feeding relationships?

_____

_____

_____

_____

_____

_____

Table 4

Team_____   Plant Species_____

Quadrat	Number of Plants Present	Approximate Area Covered
1		
2		
3		
4		
5		
6		
7		
8		
9		
10		
11		
12		

Study Questions and Additional Activities

1. View films, filmstrips, or slides on various ecosystems from representative biomes. For each ecosystem:
   a. Document the range of physical factors through a yearly cycle that appear to influence the biotic community.
   b. Note the adaptations present in response to these physical conditions.
   c. Document any symbiotic relationships observed or implied.

2. Construct a self-sustaining aquarium complete with animal, microbial, and plant life.

3. Construct a closed aquatic ecosystem in a sealed pickle jar or similar glass container. Include at least one animal and one plant. Keep in mind proper balance in cycling of materials.

4. Predict how each of the following may affect natural ecosystems:

   a. introduction of cattle for grazing purposes

   b. human recreation

   c. lumbering

   d. mining

   e. housing and/or industrial development bordering the ecosystem

5. What abiotic environmental variables might be particularly important in determining the type of ecosystem that occurs in your location? Explain.

Investigation 35 *What Changes Occur in a Temporary Pond?*

Introduction

A temporary pond is one that is created by heavy rains in wet seasons and that dries up completely in dry seasons. Temporary ponds are common in the hot, dry deserts of the southwest United States. One might expect to find temporary ponds lifeless. On the contrary, many organisms are found in temporary ponds. Where do they come from? How do their numbers and kinds change during the life of the pond? What happens to them when the ponds dry up? These are some of the questions you will attempt to answer in this investigation as you observe the organisms in your own "temporary pond."

Objectives

1. To discover and record the series of changes that take place over a period of 2 to 3 weeks in a temporary pond.

2. To create alternative hypotheses to explain these changes.

Materials

temporary pond   metric ruler
droppers   thermometer
pH paper   compound microscope
slides and coverslips   references on pond life

Procedure

1. You will try to answer two main types of questions: *Do the numbers and types of organisms vary over time? If they do, what variables seem to influence these changes?*

2. Discuss any expectations that you and your classmates have about what changes may occur. List them. Why do you think these changes will occur? These are your initial hypotheses. List them as well.

_____

_____

_____

_____

_____

_____

3. You will need to keep a careful record of conditions in your pond in order to generate meaningful conclusions about what is happening in the pond and why. Use the materials provided to gather all the data you can. Make sure to include water temperature and pH, water height in the pond, and overall appearance so that you will have a good initial reference point that you can compare with future data. **Caution: Always wash hands after handling plant and animal life.**

4. Construct a data sheet that will allow you to record possible changes in these variables if and when they occur.

5. [image] Every few days take a sample of water from the pond. Examine it closely under the microscope. Report what you see. Answer the questions in step 1 each time as well. **Caution: Use care when handling microscope and slides.**

6. Construct a graph displaying your data.

Study Questions

1. On the basis of your data, summarize the events that took place in our pond and discuss possible variables that seemed to influence them.

2. Humans manipulate succession in many ways. Cite and explain specific examples of this.

3. Why does succession result in different climax communities in different places (e.g., in Texas vs. in Central Canada or in Illinois)?

Investigation 36 *What Changes Have Occurred in Organisms Through Time?*

Introduction

While hiking through the Grand Canyon, you observe what you believe to be a fossil embedded in a layer of rock. As you continue walking, you stop occasionally and observe different fossils in different layers. This process continues until you complete your hike out of the canyon. Your observations raise many questions: *Why are different kinds of fossils found in different layers of rock? Are certain fossil forms found only in certain layers of rock? Are any fossils found in all of the layers?* You have generated more questions than can be answered during your hike. In today's exploration, you will examine various fossils collected from six of the canyon's rock layers and search for relationships, patterns, and trends in the fossil record.

Objectives

1. To discover patterns of variation in organisms through time.

2. To create and initially test alternative hypotheses to explain observed patterns in the fossil record.

3. To create alternative hypotheses about the environments in which the fossilized organisms once lived.

Materials

fossil kit containing representative fossils from six rock layers of the Grand Canyon
butcher paper
hand lens or dissecting microscope
metric ruler
pictures of the fossils from the fossil kit
cellophane tape

Procedure

1. Obtain a fossil kit for your group.

2. Spread out the fossils from the kit on a large piece of butcher paper. Place the fossils in rows with those from layer A at the top of the paper and those from layer F at the bottom.

3. Arrange the fossils to best reveal any observed trends. Draw arrows on the paper to connect possibly related fossils from one layer to the next, forming a diagram similar to a "family tree."

4. As you do this, carefully examine the fossils from each layer. Consider the following questions:

   • Are any of the fossils of an entire organism?

   _____

   _____

   • Do any of the fossils resemble an organism you might find living today? If so, what is the organism? Where does it live?

   _____

   _____

- What conditions might have been necessary for fossil formation?

_____

_____

- What might the environment in which the fossilized organism lived have been like? What evidence do you have that is consistent with your hypothesis?

_____

_____

- How do the fossils differ from one layer to the next?

_____

_____

- Are similar fossils found in more than one layer?

_____

_____

- What trends are revealed by a comparison of fossils (or lack of fossils) from one layer to the next?

_____

_____

- What alternative hypotheses can you propose to explain these trends?

_____

_____

5. Replace the fossils on the butcher paper with pictures of the fossils. Tape the pictures to the butcher paper to produce a poster of your "fossil family tree."

6. Tape your butcher paper "fossil family tree" to the board or a wall and be prepared to discuss your observed trends and those of your classmates in a class discussion.

Study Questions

1. What can be learned about living organisms by studying fossils?

_____

_____

_____

2. Some fossils found in lower layers are not found in upper layers. Give two possible explanations for this observation.

_____

_____

_____

3. Some fossils found in upper layers are not found in lower layers. Give two possible explanations for this observation.

_____

_____

_____

4. Fossils are relatively rare. Give possible reasons why this is true.

_____

_____

_____

5. Coal has been discovered at Antarctica. Give a possible explanation that could account for this occurrence.

_____

_____

_____

6. If we were to assume that organisms have not changed across time, what would the fossil record look like?

_____

_____

_____

7. What patterns would the fossil record reflect if all organisms were created at roughly the same time?

_____

_____

_____

8. Obtain a copy of the diagram at the end of this lab manual. Fill in the boxes and blanks with one causal question, hypothesis, experiment, expected result, actual result, and conclusion from this investigation. (See Reasoning Module 5 for details.)

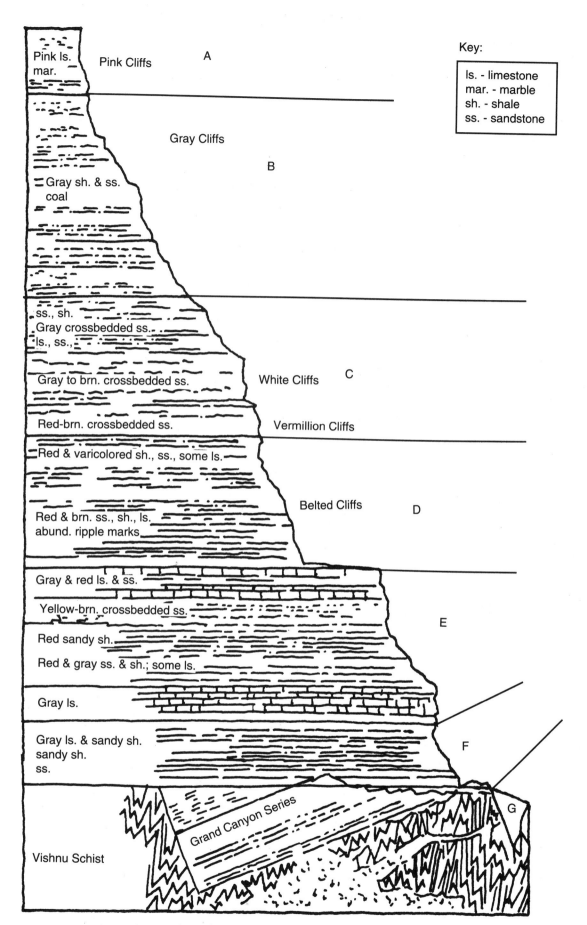

Pink ls.
mar.
Pink Cliffs   A

Gray Cliffs
B

Gray sh. & ss.
coal

ss., sh.
Gray crossbedded ss.
ls., ss.,

Gray to brn. crossbedded ss.   White Cliffs   C

Red-brn. crossbedded ss.   Vermillion Cliffs

Red & varicolored sh., ss., some ls.

Belted Cliffs   D

Red & brn. ss., sh., ls.
abund. ripple marks

Gray & red ls. & ss.

Yellow-brn. crossbedded ss.   E

Red sandy sh.

Red & gray ss. & sh.; some ls.

Gray ls.

Gray ls. & sandy sh.
sandy sh.   F
ss.

Grand Canyon Series

G

Vishnu Schist

Generalized rock layers for wall of Grand Canyon.

Investigation 37 *Have Humans Been on Earth a Long Time?*

Introduction

In the century before Darwin, the evidence that organisms changed through time was becoming increasingly convincing. In the 1600s and 1700s, the fossil record presented a complex and confusing picture to naturalists who generally believed that species did not change across time. By the late 1700s and early 1800s, however, men such as Buffon, Lamarck, and Erasmus Darwin (Charles's grandfather) argued that the fossil record implied an evolution and/or extinction of species. Buffon, for example, in his essay "Animals Common to Both Continents" (1778), suggested that the fate of the mammoth implied evolution of species:

> This species was unquestionably the largest and strongest of all quadrupeds; and, since it has disappeared, how many smaller, weaker, and less remarkable species must likewise have perished, without leaving any evidence of their past experience? How many others have undergone such changes, whether from degeneration or improvement, occasioned by the great vicissitudes of the earth and waters, the neglect or cultivation of Nature, the continued influence of favorable or hostile climates, that they are now no longer the same creatures?

Suppose you agree with Buffon that evolution has taken place. If you do, the fossil record then becomes a history of the organisms that have lived on Earth and a story of their development, proliferation, and sometimes their extinction. What organisms have lived on Earth? When did they first appear? When did they become extinct? Have humans been on Earth a relatively long time, or does the fossil record show humans to be latecomers?

Objectives

1. To generate a time line with a single scale showing a number of astronomical, geological, archeological, and historical events.

2. To discuss the relative amounts of time that various species of organisms (such as humans) have existed on Earth.

Materials

paper roll
meterstick

Procedure

1. *Have humans been on Earth a relatively long time?* To answer this question, work with a partner to construct a time line with a single scale. Place each of the events from Table 1 in its correct relative position on the line.

2. Describe how you constructed your time line.

3. Did you encounter any particular problems completing the activity? Explain.

_____

_____

4. Based on your time line, have humans been on Earth a relatively long time? Explain.

_____

_____

_____

Table 1. Important landmarks in time. Selection of time periods is somewhat arbitrary; thus, there is some overlapping. Note that columns do not represent landmarks in chronological order.

Historical	Archaelogical	Geological	Astronomical
Beginning of Dark Ages in Europe A.D. 500	Evidence of ground stone tools 10,000 B.C.	Marine algae and invertebrates abound 500 million	Probable origin of our galaxy 5.5 billion
Discovery of America A.D. 1492	Evidence of permanent dwelling 7000 B.C.	Age of Fishes 350 – 400 million	Probable origin of life 3.4 billion
Signing of the Magna Carta A.D. 1215	Evidence of woven garments 8000 B.C.	Appearance of mammals 150 million	Probable origin of Earth 4.5 billion
Birth of Christ	Evidence of domestic dogs 14,000 B.C.	Appearance of amphibians 300 million	
Fall of Egypt to Alexander 332 B.C.	Remains of a bow and arrow 11,000 B.C.	Age of Reptiles 80 – 250 million	
Beginning of Iron Age 1500 B.C.	Crude farming tools 9000 B.C.	First land plants 360 million	
Korean War A.D. 1952	Chipped stone tools 18,000 B.C.	First humans appear 1,000,000	
Writing of U.S. Constitution A.D. 1787	Evidence of humans in U.S. 17,000 B.C.	Appearance of flowering plants 160 million	
Development of Egyptian civilizations 5000 – 2000 B.C.		Modern human types appear (Cro-Magnon) 55,000	
Beginning of the Crusades A.D. 1000			

Study Questions

1. What sources of evidence have scientists used to construct sequences and dates of past events?

2. How long would your time line be (in kilometers) if your scale was set at 1 cm = 100 years?

3. The chart below shows adaptive radiation of early reptiles. Select another type of organism (e.g., mammals, birds, flowering plants, fish, algae) and use reference material to chart the adaptive radiation of their ancestors. What evidence did scientists use to construct the ancestral relationships?

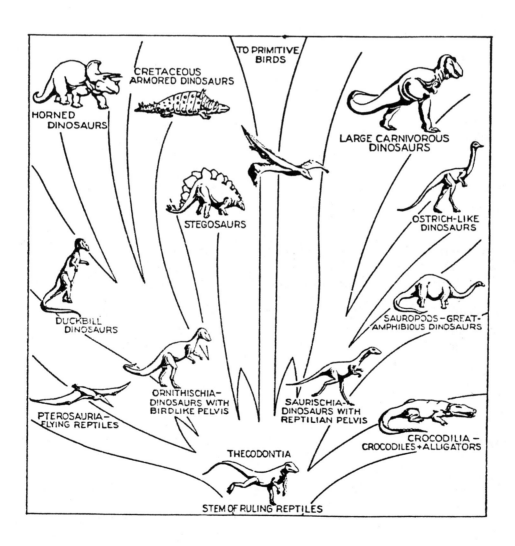

A simple family tree of the ruling reptiles. (From A. S. Romer, *The Vertebrate Story*, 4th ed., University of Chicago Press, 1959)

Investigation 38 *How Do Species Adapt to Environments?*

Introduction

Organisms have characteristics that enable them to live in particular habitats. Fish, for example, can live in water because they have gills and fins. Birds have wings and lightweight bones, so they can fly through the air. Some plants have tiny "claws" that enable them to cling to such surfaces as rocks and walls. Everywhere you look, organisms seem to be well suited to their particular set of environmental factors. How are these helpful characteristics acquired? Suppose the environment changes. Can populations of organisms acquire new characteristics to enable them to be successful under the new conditions, or must they move elsewhere to avoid extinction?

Objectives

1. To discover ways in which species change over time to become suited to particular environments.

2. To propose a mechanism for those changes.

Materials

8 pieces of patterned fabric
10 vials, each containing approximately 100 paper chips of a single color, a different color in each vial
several small containers
graph paper
film loop, *The Galapagos Tortoises*

Procedure

1. In this investigation, you will attempt to discover what happens to the characteristics of organisms within a population that is subjected to predation over a number of generations. Ideas generated from this activity may help you answer the question, *How do species adapt to environments?* Work with two or three other students.

2. To do the activity, you will play the role of a population of birds known as *Gooney birdicus* (gooney birds). Gooney birds feed on a species of mouse known as *Microtus coloriferii* (colorful mice). The role of the colorful mice will be played by the paper chips. Gooney birds are normally very hungry and always capture the first mouse they see. After the capture, they always take their prey to their nest (a small cup or vial) before they return to the hunt.

3. Begin by spreading a piece of fabric over your table. The fabric represents a natural habitat (e.g., pond, meadow, forest, cave, desert).

4. Take 10 "mice" from each of the 10 vials.

5. Spread the 100 mice at random throughout the habitat.

6. At the teacher's signal, begin capturing mice and depositing them into your nest one at a time. Your group should capture a total of 75 mice (25 mice should remain in the habitat).

7. Remove the 25 survivors by lifting and gently shaking the habitat.

8. To have the 25 survivors "reproduce," add three paper chips of the same color for each of the survivors. This new population of 100 mice consists of 25 first-generation mice and 75 second-generation mice. You may wish to make a table to help you keep track of your populations. The table should include the number of survivors and offspring of each color after each round of predation.

9. Repeat steps 5 – 8 at least two more times.

10. When you have completed your last round of predation and reproduction, display the resulting numbers of mice of each color on a bar graph with number of mice on the vertical axis and color of mice on the horizontal axis. If you have done the activity correctly, your total population should still contain 100 mice.

11. Select one member of your group to put the graph on the board to be compared with those of other groups. What patterns are revealed by the graphs? How can the results be explained?

_____

_____

_____

_____

12. Consider the following questions:
    a. What colors of prey were eaten in greatest numbers in your habitat?

    _____

    b. How does their color compare to that of their habitat?

    _____

    c. Why did you not allow those individuals that you picked up to reproduce?

    _____

    d. Did each color of mouse do equally well in each habitat? Why or why not?

    _____

    e. What might happen if the mice were all the same color at the start?

    _____

    f. What might happen if the mice were unable to reproduce at as high a rate?

    _____

    g. What might happen to the mice population if predation stopped?

    _____

    h. What might happen if mouse color were not passed from parent to offspring (if the trait were environmentally induced instead of genetically determined)?

    _____

13. View the film loop *The Galapagos Tortoises*. Use the theory of natural selection to explain how the differences in the two populations of tortoises might have come about.

_____

_____

_____

_____

Study Questions

1. During the 1920s, a population of spotted crabs was known to inhabit the white sandy beaches near a volcano on one of the Hawaiian Islands. The spotted crabs were observed to feed off plants that were cast up on the beaches by ocean waves. Occasionally, seagulls were observed capturing and

eating some of the crabs. When first observed, about 90% of the crabs were nearly completely white with only a few small black spots on their claws. About 8% of them were white with many black spots, while 2% of the crabs had so many black spots that they appeared almost completely black.

In 1930, the volcano erupted, sending a lava flow across the beach and out into the water. The lava cooled and blocked the ocean currents that had deposited the white sands on the beach. Black sands from other currents began to accumulate on the beach until within a few years the beach had become completely covered with black sand.

By the 1950s, nearly 95% of the spotted crab population was composed of crabs that were completely black. About 4% of the crabs were white with many black spots, while 1% of the crabs were white with only a few black spots.

Use the theory of natural selection to explain the change of the most frequent color of the spotted crabs from white to black. Discuss the roles of biotic potential, limiting factors, variation, and heredity.

2. Would natural selection occur more rapidly, less rapidly, or not occur at all if
   a. no variation existed in species?

   b. no changes occurred in environments?

   c. conditions of life were unlimited?

   d. characteristics were not transmitted from parent to offspring?

   e. mutations did not occur?

   f. biotic potential of species was relatively low?

3. How is the process of natural selection similar to and different from artificial selection? The inheritance of acquired characteristics?

4. Does the existence of such intermediate species as ligers and beefalos support or refute the theory of evolution? Explain.

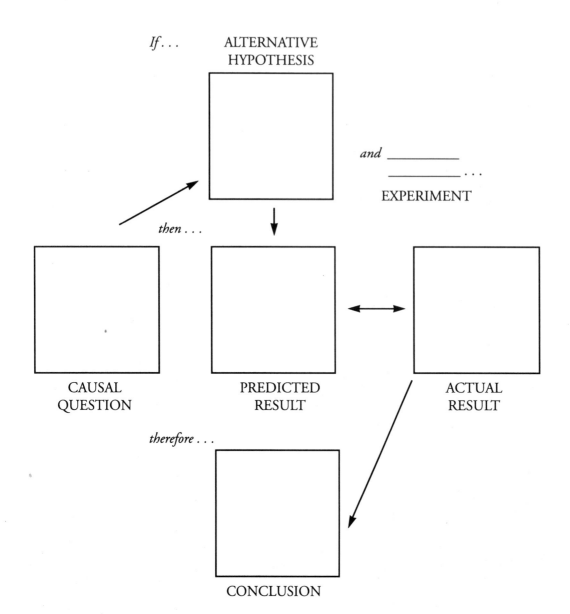

GAYLORD MG